Kunststoffe

Kunststoffe

A Collector's Guide to
German World War II Plastics

and their Markings

W. Darrin Weaver, MPA, PA-C

Schiffer Military History
Atglen, PA

Special Contributors:

Lance O. Adams
Joseph Bielik
Warren Buxton
CSM (Ret.) David W. Davis
Leon DeSpain
Bob Faris
Ulrich Finke
Dr. Georg Friedrich
D.J. Goodwin

Dr. Y. Gutierrez
Michael Heidler
Kenneth Huddle
Robert Jensen
John Kalvelage
Howard L. Kelley
Dieter H. Marschall
Cipriano Moreno
CPT (Ret.) Mack A.

Pattarozzi
Paul Seibold
R. Blake Stevens
Dr. Geoffrey Sturgess
Paul Viens
Hamil Ward
Jason C. Weaver
Margaret J. Weaver
Dr. David K. Yelton

Contributing Companies and Institutions
Bakelit-Museum, Altes Amtshaus, Kierspe, Germany
Bundesarchiv, Koblenz, Germany
Collector Grade Publications, Cobourg, Ontario
Deutsches Patent- und Markenamt, München, Germany
Manion's International Auction House, Inc., Kansas City, Kansas
Nordrhein-Westfalen-Stiftung Naturschutz, Heimat- und Kulturpflege,
    Düsseldorf, Germany
United States Army
United States Army Ordnance Museum, Aberdeen Proving Ground, Maryland
United States Department of Veteran's Affairs
United States Library of Congress, Washington, D.C.
United States National Archives and Records Administration, College Park, Maryland
United States Patent and Trademark Office, Department of Commerce, Washington D.C.

Book Design by Ian Robertson.
Copyright © 2008 by W. Darrin Weaver, MPA, PA-C.
Library of Congress Control Number: 2007937645

All rights reserved. No part of this work may be reproduced or used in any forms or by any means – graphic, electronic or mechanical, including photocopying or information storage and retrieval systems – without written permission from the copyright holder.

Printed in China
ISBN: 978-0-7643-2923-4

We are interested in hearing from authors with book ideas on related topics.

Published by Schiffer Publishing Ltd.
4880 Lower Valley Road
Atglen, PA 19310
Phone: (610) 593-1777
FAX: (610) 593-2002
E-mail: Info@schifferbooks.com.
Visit our web site at: www.schifferbooks.com
Please write for a free catalog.
This book may be purchased from the publisher.
Please include $3.95 postage.
Try your bookstore first.

In Europe, Schiffer books are distributed by:
Bushwood Books
6 Marksbury Avenue
Kew Gardens
Surrey TW9 4JF, England
Phone: 44 (0) 20 8392-8585
FAX: 44 (0) 20 8392-9876
E-mail: Info@bushwoodbooks.co.uk.
Visit our website at: www.bushwoodbooks.co.uk
Free postage in the UK. Europe: air mail at cost.
Try your bookstore first.

Contents

	Introduction	6
Chapter 1:	Brief History of Early Plastics Developments	8
Chapter 2:	German Military Use of Polymer Plastics	16
Chapter 3:	Thermoplastic and Thermoset Materials	34
Chapter 4:	Staatliches Materialprüfungsamt Berlin-Dahlem	36
Chapter 5:	The Manufacturing Process and Constituents	39
Chapter 6:	Types and Identification of Pressing Compositions	47
Chapter 7:	Quality Control	52
Chapter 8:	Other Methods of Differentiating Various Materials	53
Chapter 9:	Examples of Manufacturers	60
Chapter 10:	Manufacturers' Logos, Markings and Codes	62
Chapter 11:	Selected Notes on Common Wartime Products	90
Chapter 12:	Selected Topics and Interesting Military Products Made during WWII	103
Chapter 13:	Allied Appraisal of the Wartime German Plastic Industry	109
Chapter 14:	Post-War Miscellany	111
Appendix I:	Material Classification Reference	113
Appendix II:	Manufacturers by Observed Trademarks/Logos	114
Appendix III:	Abbreviations and Company Initials	123
Appendix IV:	Alphabetical Listing of Kunststoff Manufacturers	128
Appendix V:	Manufacturers by Two-Digit Number Designation	138
Appendix VI:	Manufacturers by Letter-Number Designation	142
Appendix VII:	Manufacturers by Number-Letter Designation	148
Appendix VIII:	Military Letter Codes Assigned to Kunststoffe Manufacturers	152
	Glossary and Abbreviations	165
	Bibliography	169
	Notes	172

Introduction

The primary purpose of compiling this short text is to provide the militaria, ordnance, and small arms researcher, collector, or enthusiast a historical overview and easy to use reference covering prominent wartime plastics manufacturers, markings, logos, codes, and material identification techniques of wartime products. In conjunction with this, I have provided the reader a brief summary of the German pre-war and wartime plastics industry, its major end products, as well as a short familiarization of the constituents and manufacturing process. The goal of this project was not to author a "Chemistry 101" textbook, and although my academic degrees are thoroughly based in the medical sciences, in regard to this reference, I am a collector of artifacts and a compiler of information—not a chemist.

I do find both the ornate and the utilitarian forms of early and military polymer plastics interesting, especially those with high proportions of wooden filler, or those with "marbling." It was, however, a certain controversy in my first firearms book; that is, the actual material used to make K43 rifle hand guards marked "Durofol" that certainly piqued my interest in the WWII German plastics industry. I simply could find no English language references specifically tailored to the collector or militaria enthusiast at the time, and other books contained only an anecdote or two concerning the matter, and this perhaps inspired me to dig a little deeper into the German wartime plastics industry.

I have heavily excerpted and referred throughout the text to two very important German trade publications of the era: first, the 1940 *Bekanntmachung über Kunstharz-Preßmassen für typisierte und überwachte Preßstoffe* (Announcement Concerning Synthetic Resin Molding Compounds for Standardized and Inspected Moldings) tabulated

Introduction

by the *Staatliches Materialprüfungsamt Berlin-Dahlem* (State Material Testing Office in Berlin-Dahlem), and published in the 1940 trade journal *Kunststoffe*, kindly provided by the esteemed small arms collector Dr. Geoffrey Sturgess. Second, *Werkstoff Ratgeber* (Materials Advisor) by *Dr.-Ing.* Herwarth v. Renesse, published in 1943, and provided by the noted collector and exceptional author, Dieter H. Marschall. Both of these documents were painstakingly translated for me by the gifted scholar Paul Seibold, without which this reference would not have been possible.

I wish to extend my gratitude to all of the contributors who helped with this project, and especially those that proof read and corrected my work. I would also like to thank my wife and children for tolerating and encouraging my research trips and late night efforts, as well as CPT (Ret.) Mack Pattarozzi and Joseph Bielik, my longtime friends, colleagues, and traveling companions for their never-ending support. So too, my talented brother, Jason C. Weaver, deserves special mention for assisting with yet another excellent cover concept.

A number of German military letter code references have already been published, and the intent here was not to duplicate these efforts. However, I have listed these codes where practical and/or relevant to the subject matter. I wish to extend my thanks to two other individuals, both friends and colleagues: the German author Michael Heidler, and noted American collector and authority, Kenneth Huddle, who were both quite helpful in this regard, and also proofed the text.

Of special note, many of the names and products discussed in this work are proprietary and protected by international trademark laws and agreements. Some have become colloquial generics over time, the status of others is unclear, and many of the original trademark permits have expired. Trademarks and logos have been reproduced solely for reference and product identification, and no infringement is intended. Most images of trademarks were obtained from public records on file at either the United States Patent and Trademark Office or the Deutsches Patent- und Markenamt. Those that could not be found there were recreated digitally by myself.

Images used and reproduced by permission are credited in the text. Most of the images of the machinery used in wartime production were taken from the reprints of original U.S. and British Army reports on hand at the U.S. Library of Congress, and while the quality of each is poor, they are in some cases the only contemporary illustrations available. The remainder of the images not directly credited were photographed by myself of items in my personal collection, or were obtained from my collection of wartime press and periodical images, all of which are thought to be public domain.

I thank you for purchasing this book, and I hope that you find it interesting and useful.

W. Darrin Weaver
2006

1

Brief History of Early Plastics Developments

Today, nearly everything can be and is made of plastic. Modern consumers, and yes, collectors simply take this for granted, so ingrained are the various materials in our daily lives. But, what frequently is not appreciated is just how truly revolutionary the invention of synthetic polymer plastics was over a century ago.

The word "plastic" is derived from the Latin *"plasticus,"* or something that is "...capable of being formed or shaped."[1] The German word for plastic is *Kunststoff*, or *Kunststoffe* in the plural, hence the title of this work.

"Plastics" in a nutshell are "polymers," or as defined, "...something made of many units," or a "chain." Each link of the chain is a basic molecular unit (a monomer), usually carbon, hydrogen, oxygen, and/or silicon. These basic, small molecules are joined together into these longer chain-like molecules via the process known as polymerization.

Mankind had put naturally occurring polymers such as tar, shell, horn, amber, and latex to good use since the beginning of time to make various items. These natural polymers certainly could be considered "plastic," but the term over time has evolved a meaning generally representing those polymers either manipulated or created synthetically by human manipulation. It was not until the 1800s, though, that chemists began experimenting in earnest with, and modifying these naturally occurring polymers into such goods as Vulcanized latex rubber and Celluloid.

Parkesine was actually the first polymer plastic, and was announced in 1862 by Alexander Parkes (1813-1890) in London at the Great International Exhibition (British Patent 1313 1865). This cellulose based material could be heated and formed, but was not a financial success, partly because of the attempt to keep the price of the material under a

Chapter 1: Brief History of Early Plastics Developments

Shilling per pound. Celluloid, invented by John Wesley Hyatt (1837-1920), and patented in 1869 (U.S. patent number 105338 1870), came next, and was the first commercially successful plastic. Celluloid was also organic in origin, made from cellulose-containing vegetable matter (typically cotton treated with nitric or sulphuric acids) and pulverized gum camphor. Celluloid was often used as a less expensive replacement for more traditional and expensive natural polymers such as horn, ivory, or tortoise shell, used for instance in the manufacture of combs, brush handles, jewelry, knobs, etc. Celluloid was indeed useful, but was flammable, and had many other disadvantages from a manufacturing standpoint.

A German, Dr. Adolph Spippeler, next developed another organic polymer (Casein-formaldehyde) by combining the protein from sour milk whey with formaldehyde. This light colored material, later marketed under the trade name "Galalith," showed much promise, but its proprietors failed to get the product established in the market before another material took the industrialized world by storm. The process of producing Casein on a large scale was eventually perfected, commercialized, and used some years later to make products such as billiard balls, buttons, handles, and other items, but it too had limitations and manufacturing disadvantages.

Despite the development of both Celluloid and Casein, most items made prior to the turn of the last century were still constructed from traditional materials: iron, steel, wood, bone, shell, etc. So that, if one wanted to manufacture, for instance, a table clock, each

Fig. 1: The interior of a small, well equipped turn of the last century Celluloid processing plant, that of B. Georgi, Offenbach am Main, Germany.

component would have to be cast, forged, machined or shaped by hand, assembled by craftsmen, and usually housed in some sort of wood furniture. This same housing would have to be shaped and then sanded, again by workers, and finished.

What this all amounted to was a fairly inefficient and time consuming process of manufacturing that was highly dependent upon skilled and expensive labor.

To return to the hypothetical table clock example: Once it was completed, the wooden housing would be stained and protected, usually with some sort of varnish. Varnishes are substances that transform into polymers when drying. The most frequently used type was a natural polymer called Shellac, derived from resins secreted and left on trees by the South Asian *Laccifer Lacca* beetle. The resins were collected by the local populace, and then processed and filtered by slow and relatively inefficient methods. Shellac was, however, used for generations as a coating and preservative on all sorts of wood products.

Then the phenomenal rise of the electrical industry in the late 1800s created an instant need for efficient insulating materials. Glass and porcelain were terrific insulating materials, but were heavy, and not particularly cheap. Manufacturers soon discovered that fairly light

Fig. 2: Some examples of the sorts of products that could be made from Celluloid and ordered by catalog from the same company. Celluloid processing, however, did not lend itself well to what we now know as "mass production," and had other drawbacks.

Chapter 1: Brief History of Early Plastics Developments

and efficient insulators could be made out of Shellac and pressed paper, and the polymer could also be applied directly to conductive surfaces. Soon the supply of Shellac simply could not keep up with demand, and dozens of Western firms and chemists began working toward some sort of synthetic substitute.

One man sort of stumbled upon the answer earlier in 1872. Adolf von Bayer (1835-1917), the noted German chemist (responsible for aspirin, heroin, and methadone, among others), noted that a persistent insoluble residue occurred in his glassware after experiments that involved phenol (carbolic acid) and formaldehyde. Unfortunately for him (and the world), Bayer failed to really recognize the potential utility, or exploit the full potential the residue might achieve with further experimentation and processing.

In 1889 a Belgian chemist experimenting with photographic chemicals and processes, Leo Henrik Baekeland (1863-1944), moved from his native country to the United States. Baekeland then set about making a name for himself in the 1890s by inventing and marketing the photographic paper Velox, which allowed photographers to develop their film with artificial light instead of sunshine. This was a revolutionary development in

Fig. 3: The ingenious Leo H. Baekeland (1863-1944). *US NARA*.

photographic processing, such that in 1899 Eastman Kodak bought all rights to Velox from Baekeland for the astonishing sum of $1,000,000. This windfall allowed Baekeland the financial freedom to set up a lab and work on other projects on the grounds of his Yonkers, New York, home.

Baekeland originally set out to develop a replacement for Shellac as a coating, but he and others soon recognized the need and potential market for some sort of a truly synthetic replacement for Shellac in the manufacture of electrical insulators. Baekeland studied the earlier work of notable chemists such as Bayer, Blumer, and Kleeburg, and concentrated his efforts on phenol and aldehyde chemicals. Actually, an inventor named A. Smith had filed and was granted the first patent for phenolic resin as substitute for hardened rubber in 1899 (German Patent 112 685). Baekeland's experiments began in earnest in 1904. Through his work Baekeland further refined the reaction, and had the vision to see its broader potential.

Fig. 4: Leo Baekeland's "Bakelizer." US NARA.

Perhaps more importantly, he began development of an innovative "heat and pressure" technique that would eventually make possible the successful commercialization of the resin. In 1907, after years of research and experimentation, he came up with the world's first truly successful totally synthetic polymer plastic, which he called "Bakelite."

Bakelite (actually, polyoxybenzylmethylen-glycolanhydride) was the end product of a controlled polymerization reaction involving heat and pressure between phenol (a solvent distilled from coal tar) and formaldehyde (an embalming fluid from methyl alcohol). When the two were heated (with a base or an acid) they formed a substance similar to Shellac. Additional heat and time created a thicker resin. Upon further heating, to about 100 C degrees and under pressure, the resin would form an exact near-colorless replica of any vessel or mold that contained it.[2] One of the keys to this process was a type of pressure cooker-type apparatus the chemist called the "Bakelizer."

Through further testing and experimentation, Baekeland discovered that his new material would not boil, melt, or scorch, and perhaps most importantly, had excellent electrical insulating properties. It also did not shatter, crack easily, or crease, and was essentially unaffected by sunlight, moisture, or the salinity of seawater. Pigments and powders could be added to the mix by various techniques to simulate bone, ivory, shell, wood, or marble. In addition, Bakelite would not dissolve in acids or solvents, could be combined with "fillers" (such as wood pulp, fibers, and other substances) for strength, and once formed and set, the material would never change.

Baekeland patented his revolutionary process (U.S. Patent 942699, British Patent 1921 1909, German Patent 233 803), and announced his discovery in 1909 at the New York chapter of the American Chemical Society. A shrewd businessman, Baekeland traveled to Germany and founded the Bakelite Gesellschaft m.b.H in Erkner, outside of Berlin, in 1910, by incorporating with two German firms, Rutgerwerke A.G. and Knoll & Company. A few months later, back in his adopted country, Baekeland then created the General Bakelite Corporation, and set up shop in Perth Amboy, New Jersey, to manufacture and market the material from North America.

For the first time a consumer product, such as a hair comb, which had previously taken the skills of an expensive laborer to carve or produce from bone, horn, or inefficient celluloid, could be mixed and molded by the thousands in a fraction of the time. The mass production potential was enormous, and needless to state, Bakelite was a near instant success. Before long the material dominated the electrical insulator industry, and was quickly applied to produce a cornucopia of both commercial and military products.

Over the next few years, as the success of Bakelite grew, Leo Baekeland continued to refine the product, and spent much of his time filing patents and fighting off imitators, knock offs, and patent infringements. Indeed, over the span of his career, Baekeland filed over one hundred domestic and foreign patents, and was a member, and served one year as chairman, of the Committee on Patents of the National Research Council.

The financial success of Bakelite and Baekeland's spirited defense of his proprietary rights did much to accelerate the creation of other synthetic materials. The same year that Casein actually made it into production Rayon was patented (1913). Other revolutionary "wonder" materials soon followed. In 1923 Frits Pollack produced Catalin (Urea-formaldehyde). Then, Cellophane (1925) and "PVC," or Polyvinyl Chloride (1926), came next.

Leo Baekeland's original patent expired in 1927. This allowed thousands of companies all over the world to freely produce goods using his original method, or modifications of the same. Before long, the global plastics industry simply exploded with new products, techniques, and materials.

Paralleling the rapid growth in the use of the phenolics, further revolutionary materials were patented during this era. Polystyrene was developed in 1930, Neoprene (1932), Melamine (1933), Polyvinylidene Chloride (better known as Saran, 1933), Polyethylene (1933), Nylon (1934), Plexiglas (1935), Teflon (1938), and Formica (1940) all followed in turn.

Few of these later materials saw widespread use in WWII. Once a material was announced or patented, the inventor usually found it necessary to scurry about for financial capital. At the same time, processes had to be refined and matched to appropriate applications. So too, raw material and transportation networks had to be developed and actual manufacturing lines set up. Thus, there usually was a gap of several years, or even decades between the time a material was announced or patented and the time its end products were available commercially.

So, by the eve of WWII it was still Leo Baekeland's original phenolic polymer plastic, in both trademarked and generic forms, that dominated the worldwide plastics industry.[3]

Fig 5 (Opposite): The front page of Leo Baekeland's ground breaking 1909 US patent, number 942,699. *US Patent and Trademark Office.*

UNITED STATES PATENT OFFICE.

LEO H. BAEKELAND, OF YONKERS, NEW YORK.

METHOD OF MAKING INSOLUBLE PRODUCTS OF PHENOL AND FORMALDEHYDE.

942,699.

No Drawing.

Specification of Letters Patent.

Application filed July 13, 1907. Serial No. 383,684.

Patented Dec. 7, 1909.

To all whom it may concern:

Be it known that I, LEO H. BAEKELAND, a citizen of the United States, residing at Snug Rock, Harmony Park, Yonkers, in the county of Westchester and State of New York, have invented certain new and useful Improvements in Methods of Making Insoluble Condensation Products of Phenols and Formaldehyde, of which the following is a specification.

In my prior application Ser. No. 358,156, filed February 18, 1907, I have described and claimed a method of indurating fibrous or cellular materials which consists in impregnating or mixing them with a phenolic body and formaldehyde, and causing the same to react within the body of the material to yield an insoluble indurating condensation product, the reaction being accelerated if desired by the use of heat or condensing agents. In the course of this reaction considerable quantities of water are produced, and a drying operation is resorted to to expel it.

The present invention relates to the production of hard, insoluble and infusible condensation products of phenols and formaldehyde.

In practicing the invention I react upon a phenolic body with formaldehyde to obtain a reaction product which is capable of transformation by heat into an insoluble and infusible body, and then convert this reaction product, either alone or compounded with a suitable filling material, into such insoluble and infusible body by the combined action of heat and pressure. Preferably the water produced during the reaction or added with the reacting bodies is separated before hardening the reaction product. By proceeding in this manner a more complete control of the reaction is secured and other important advantages are attained as hereinafter set forth.

If a mixture of phenol or its homologues and formaldehyde or its polymers be heated, alone or in presence of catalytic or condensing agents, the formaldehyde being present in about the molecular proportion required for the reaction or in excess thereof, that is to say, approximately equal volumes of commercial phenol or cresylic acid and commercial formaldehyde, these bodies react upon each other and yield a product consisting of two liquids which will separate or stratify on standing. The lighter or supernatant liquid is an aqueous solution, which contains the water resulting from the reaction or added with the reagents, whereas the heavier liquid is oily or viscous in character and contains the first products of chemical condensation or dehydration. The liquids are readily separated, and the aqueous solution may be rejected or the water may be eliminated by evaporation. The oily liquid obtained as above described is found to be soluble in or miscible with alcohol, acetone, phenol and similar solvents or mixtures of the same. This oily liquid may be further submitted to heat on a water- or steam-bath so as to thicken it slightly and to drive off any water which might still be mixed with it. If the reaction be permitted to proceed further the condensation product may acquire a more viscous character, becoming gelatinous, or semi-plastic in consistence. This modification of the product is insoluble or incompletely soluble in alcohol but soluble or partially soluble in acetone or in a mixture of acetone and alcohol. The condensation product having either the oily or semi-plastic character may be subjected to further treatment as hereinafter described. By heating the said condensation product it is found to be transformed into a hard body, unaffected by moisture, insoluble in alcohol and acetone, infusible, and resistant to acids, alkalies and almost all ordinary reagents. This product is found to be suitable for many purposes, and may be employed either alone or in admixture with other solid, semi-liquid or liquid materials, as for instance asbestos fiber, wood fiber, other fibrous or cellular materials, rubber, casein, lamp black, mica, mineral powders as zinc oxid, barium sulfate, etc., pigments, dyes, nitrocellulose, abrasive materials, lime, sulfate of calcium, graphite, cement, powdered horn or bone, pumice stone, talcum, starch, colophonium, resins or gums, slate dust, etc., in accordance with the particular uses for which it is intended, and in much the same manner as india rubber is compounded with the above-named and other materials to yield various valuable products. In compounding the condensation or dehydration product in this manner the desired materials are mixed with the same before submitting it to the final hardening operation below described.

2

German Military Use of Polymer Plastics

Germany, like other soon-to-be combatant nations, was quick to apply new plastics technology in military applications. Casein and Celluloid had seen limited use by the Germans in WWI to form such items as binocular eyepieces and other small components, but fell from favor when the use of phenolic polymer plastics really came of age after

Fig. 6: A multi-piece game made of phenolic *Kunststoff* that the author speculates was similar to the more famous "Scrabble" familiar to most readers.

Bakelite G.m.b.H. was established at Erkner. This use of phenolics accelerated in the 1920-1930s, which should have come as no surprise, as Germany possessed considerable coal reserves, the raw material from which phenol is distilled. However, the trend simply exploded after the original Bakelite patent expired in 1927, and numerous firms cropped up all over the country using Baekeland's formulas and techniques, or variations and modifications of the same.

Some German civilian and industrial applications from the period as observed during the author's research include major end products such as: automobile, motorcycle, and bicycle components; promotional ashtrays; buttons; cameras; clocks; containers; dishes; electrical insulators/equipment; electric horns; film canisters/rolls; game pieces; gauges; gears; scientific and measuring instruments; kitchen utensils; knife and razor handles;

Fig. 7: Finely mottled phenolic cigarette box made by H. Römmler Aktiengesellschaft, Spremberg Nd.-Lausitz, is another example of the thousands of products made by the *Kunststoffe* industry for the commercial market prior to the war.

Fig. 8: A sampling of the many types of "safety" razors made almost entirely of phenolic resins, widely available to both civilian and military personnel.

knobs; hair dryers; jewelry; political "tinnies"; radios; switches; toys; telephone housings; and typewriter casings and keys; and even full length skis, to name but a few.

After 1927, many German companies subsequently developed their own proprietary names or registered commercial brands for identifying and marketing the phenolic and other polymer plastics they produced. Indeed, by 1938 there were over sixty such trade names known, to include: Agfenit, Alberit, Alusil, Ambroin, Bakelite, Bebrit, Bezeg, Columbus, Deligna, Deurohlit, Durax, Dynal, Elgesit, Escolith, Eshalit, Faturan Fermit, Fibresinol, Formolit, Friwocit, Graconit, Hares, Helosit, Heliowatt, Hercules, Isolierpanzer, Isolierpreßtoff, Jaroplast, Kerit, Licolit, Mendolith, Mixit, Neoresit, Norit, Pertinit, Phenoplast, Pollopas, Preßtoff, Ralotext, Reicolit, Resiform, Resinol, Resistan, Resopal, Rö, Spritzstoff, Taumalit, Teasit, Teka-Preßtoff, Tenacit, Thormalan, Trolitan, Tundes, Wecolit, Wellit, Werkstoff, Zeton, and Zeterit, among others.[4,5]

In 1933 Adolf Hitler and the German National Socialists seized power, and a short time later announced that Germany was disregarding the Treaty of Versailles and would rearm. Of course, the success and potential military applications of phenolic polymer plastics was not lost on Germany's revitalizing military, nor its ordnance and small arms designers and engineers, either. Soon, the German military-industrial machine was in full expansion,

Chapter 2: German Military Use of Polymer Plastics

Fig. 9: An assortment of personal items confiscated from German POWs. The individual soldier relied heavily on items made from *Kunststoffe*, including compasses, knife handles, buttons, razors, whistles, oilers, and various containers, many of which are depicted here. *US NARA*.

Fig. 10: A set of 1941 dated tent pegs made by Gebr. Spindler Betr.-Kom.-Ges., KG, Köppelsdorf.

Fig. 11: Rugged *Heer* marching compass with at least the black phenolic body made by F.G. Zieger of Rosswein.

Fig. 12: Typical soldier's canteen, *sans* cup. A majority of *Wehrmacht* canteen caps and a vast number of cups were made from *Kunststoff*.

Chapter 2: German Military Use of Polymer Plastics

Fig. 13: Each soldier in the German Heer was authorized approximately 72 grams of some sort of dietary fat per day, which could include butter, margarine, lard or "*Schmalz*" (an onion/rendered animal fat concoction), which was spread on bread and consumed or used for cooking. This is a typical *Fettbüchse*, or "fat container," issued to every soldier as part of his personal equipment, and could hold about three days, or 220 grams, ration. These "butter dishes" are found in a wide array of colors, such as off-white, orange, green, brown, or black. This particular example is actually made from urea-formaldehyde.

Fig. 14: Binoculars were an essential accessory for key leaders, spotters, and snipers. A solemn-looking *Waffen-SS* member in this wartime press photograph scans his sector with 6x30 binoculars at the ready. The wartime German optical industry consumed tens of thousands of tons of phenolic material to produce eyepieces, components, protective covers, and waterproof cases. *US NARA*.

Chapter 2: German Military Use of Polymer Plastics 23

Fig. 15: Hard, water-tight phenolic cases were developed for the military to protect sensitive optical equipment. Binocular cases were made by a number of manufacturers, and were made in a number of hues, such as red-brown (above) and near black (below), both of which are typical of the type.

Fig. 16: Another *Waffen-SS* NCO is shown with a phenolic binocular case hung over his shoulder as he examines a damaged tank in this wartime press photo.

Chapter 2: German Military Use of Polymer Plastics

Fig. 17: A *Heer* communications specialist checks the line. Manufacturers of communications and electronic equipment were probably the heaviest users of phenolics prior to and during the war. The field phone and handset shown in use in this press photograph were contained in phenolic housings, and many of the internal components were made from or insulated by phenolics.

Fig. 18: A 1939 dated field phone in phenolic housing, and composed of many synthetic polymer components.

Fig 19: A hardy piece of wartime communications gear, the HLS A 544, with many components, as well as the exterior housing manufactured from *Kunststoff*. US NARA.

Chapter 2: German Military Use of Polymer Plastics

Fig. 20: Rugged set of *Wehrmacht* wire cutters, insulated with layered pressed paper bonded together with phenolic glue, and capped with sturdy phenolic end pieces.

and increasingly incorporated the use of phenolic polymer plastics to simplify and hasten ordnance and small arms production.

In the end, the *Wehrmacht* used hundreds of thousands of tons of phenolic polymer plastics (and to a lesser extent, others such as urea-formaldehydes and melamine) before, during, and through the end of WWII. Some prominent examples of common military products identified during the author's research include: automotive, aviation, and armored vehicle components; binoculars, binocular cases, and accessories; belt buckles; butter dishes; buttons; canteen caps and cups; carbide lamps; cigarette lighters; compass housings; containers; ear phones; field phones; flashlight bodies; fuses; fuse holders; instrument housings; mess/food containers; optical equipment; razors; signaling equipment; telegraph equipment; and individual tent stakes, to name but a few.

In regard specifically to ordnance and small arms: bayonet and knife handles, fuses and fuse components, fuse holders, G41 rifle hand guards, magazine bottoms, MG34 butt stocks, small arms oilers, pistol/MP/StG grip plates, powder/cartridge containers, rifle grenade driving bands, and even complete rifle stocks were all made of phenolic and other polymer plastics, or used components made of the same in their assembly.

Fig. 21: A well-equipped NCO, as shown in this familiar wartime press photograph. His MP40 submachinegun was equipped with phenolic grip plates, and the bulk of the lower receiver was housed in it as well. In addition, much of a soldier's personal items were made of *Kunststoffe*. It can be assumed that, in addition to the binoculars and SMG shown, he also carries a "butter" dish, weapons cleaning kit with phenolic oiler, canteen with *Kunststoff* cup and cap, and perhaps a razor, cigarette case, and a cigarette lighter made of the material. This, in addition to the many buttons on his trousers and underclothes, were frequently manufactured from phenol-formaldehyde. Indeed, the very survival of the individual soldier depended in no small part on the wartime plastics industry.

Chapter 2: German Military Use of Polymer Plastics

Fig. 22: SG 84/98 bayonets, issued into the 1930s for use with the K98 rifle, were stocked with wooden grip plates. Cutting, shaping, sanding, and final fitting of these was time consuming (top). Phenolic grip plates were developed that could be turned out in the tens of thousands, and either held in place by screws (second from top), or by permanent rivets (second from bottom). The SG42 was developed to replace the SG 84/98, and was fitted with a different pattern of *Kunststoff* grip plates, again rivetted in place.

Fig. 23: Two bayonets with different grip plate hues.

Fig. 24: Officers and NCOs of the Heer are practicing with a P38 pistol in this wartime press photograph. Both the pistol and the binoculars in this photo were manufactured with phenolic components, i.e. grip plates, binocular eyepieces, covers, and protective containers.

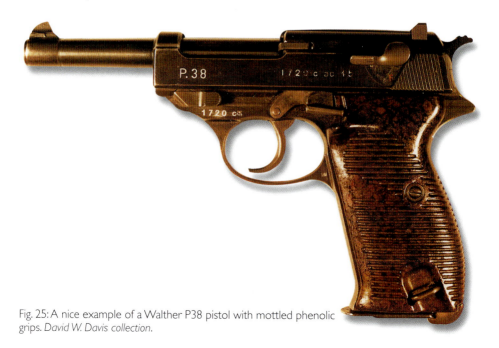

Fig. 25: A nice example of a Walther P38 pistol with mottled phenolic grips. *David W. Davis collection.*

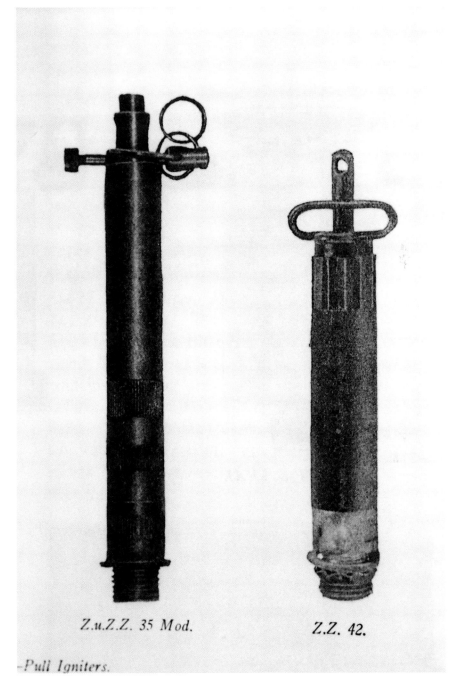

Z.u.Z.Z. 35 Mod.　　　　Z.Z. 42.

—Pull Igniters.

Fig. 26: Various items related to explosives were made from phenolics during the war because of the material's affordability and insulating qualities, and they could not generate accidental sparking. One common item made from phenolic resin was the "pencil" igniter. *From US Army TM-E-30-451.*

Fig. 27: Phenolics were just about perfect for any sort of electrical housings, an example of which is shown here, a wartime German demolitions "twist" igniter. *From US Army TM-E-30-451.*

Chapter 2: German Military Use of Polymer Plastics

Fig. 28: A *Zt.Z.f. SpBü. 37* explosives timer made up of a clockwork housed in a *Kunststoff* body. From US Army TM-E-30-451.

3

Thermoplastic and Thermoset Materials

So what exactly is a wartime polymer plastic? "Bakelite" as a term has long been used inaccurately by collectors as a generic for any WWII German item or component made of "plastic." But the use of this name—a proprietary brand of Bakelite Ltd. in England, Bakelite Corp. (and later Union Carbide, and still later Georgia Pacific) in the U.S., and Bakelite A.G. in Germany—is not really correct.[6] As will be discussed, the phenolics are but one (albeit the first and most important) group of plastics used during the era. So too, many items produced during the war were not made from phenolic based material at all, but rather were produced from entirely different materials.

Polymer plastics, or "*Kunststoffe*," are made up of three main sub-types: urethanes, thermoplastics, and thermosets.

Urethanes are not particularly relevant to this discussion, as their true potential was not realized and exploited until well after the war. Thermoplastics include early nitrates (such as Celluloid), and later developments, such as acetates, butyrates, ethyls, vinyls, acrylics, polystyrenes, nylons, and polyethylenes. Generally, thermoplastics are made from non-curable molding compounds, will soften when heated, and can be recycled and/or remolded with the right processes and techniques. Like the urethanes, thermoplastics (with the exception of Celluloid, known in German as both *Zellhorn* or *Zelluloid*) were not exploited to any great extent prior to or during World War II, and thus will not be discussed here in any great depth. Thermoplastics did, though, see accelerated use after the war, and nearly everything today is made from them.

Thermosets, on the other hand, are much more relevant to the WWII German militaria, ordnance, and firearms collector, and include the phenolics, ureas, melamines, and polyesters. Nearly all of the belligerent forces during the war used thermosets to some degree in the manufacture of military goods.

The first successful synthetic proprietary polymer plastic of this category was of course Bakelite and its imitators, as has been discussed. These phenol-formaldehyde plastics, known initially as *"Phenolharz-Kunststoffe"* in German, and the other thermosets are curable molding compounds. The advantages of thermosets are that they are of course moldable into a myriad of shapes and forms, are comparatively light in weight, have excellent insulating qualities, are cheap to produce, and are less susceptible to expansion and contraction than wood or metal.

Thermosets do, however, have some disadvantages. Phenolics, ureas, and melamines in their finished state can be brittle, may be subject to decay (due mostly to the organic filler), and cannot be recycled. Once formed and cured (the process of cross-linking) they are "set," and when heated or put to a flame they do not melt or soften, but rather burn or scorch.

4
Staatliches Materialprüfungsamt Berlin-Dahlem

Industrial standards are vital, and of ever increasing use today. Indeed, the reader is likely already familiar with, or has seen the abbreviations of such modern entities as the ASA (American Standards Association). Most of the industrialized economies of the world established their own national associations which helped to ensure that a product made in one location was of the same composition and/or quality as the same product made on the other side of a country (and these days, on the other side of the world, through the ISO, or International Standards Organization). Germany was no different, and neither was its plastics industry.

In 1924 the V.D.E., or *Verband Deutscher Elektrotechniker* (German Federation of Electrical Engineers), first attempted to present guidelines for plastics production and material classifications, as many of its members either used or produced plastics for insulators. In conjunction and paralleling this, the German plastics industry itself became well established, and the various production companies were organized into a trade association, or guild, similar to those found in other industries all over the world. This group was called the *Technische Vereinigung der Hersteller und Verarbeiter typisierter Kunststoff-Formmassen e.V.*, or the Technical Association of the Processing Plants and Manufacturers of Standardized Plastics (many of whom also belonged to the German Federation of Electrical Engineers). The headquarters of the group was responsible for keeping track of the individual manufacturers, developing and publishing industry standards and norms (DIN, or *Deutsche Industrie Norm*), and was in existence until the late 1990s.

The DIN for each classification was modified periodically and updated as new techniques and processes were developed, and expectations of quality changed. The DIN that covered the German plastics sector was DIN 7702 (The growth of the global economy and the establishment of the rise of the International Standards Organization [ISO] in the late 1990s essentially rendered all national standards associations obsolete).

The *Staatliches Materialprüfungsamt*, or State Materials Testing Office, located in Dahlem, near Berlin (now a district of Berlin itself), abbreviated MPD, was established in 1936 (some sources state that the agency originated as early as 1871), and was the responsible authority tasked with monitoring, inspecting, and enforcing the German industrial standards, both DIN 7702 and DIN 7708. The agency was led by President-Professor Sieber, and Professor Kindsher served as both Vice-President and chief of the "organic" section, which included plastics, rubber, paints, artificial fibers, and glues. Prior to the war, the MPD had a staff numbering some 450 personnel, of whom 320 were qualified technicians or chemists. 140 such employees served under Kindsher at the MPD's height, but by the end of the war, the entire agency numbered only 60, the remainder having been drafted into the *Wehrmacht* due to military personnel shortages.[7]

The MPD Marking

Each article was supposed to be marked, or have a marking incorporated in the molding process to demonstrate which company made it, exactly what it was made of, and that it was produced to DIN standards. A marking composed of the stylized combination of

Fig. 29: The stylized MPD marking as prescribed by the *Staatliches Materialprüfungsamt*. The presence of the marking demonstrated that the product was produced in accordance with DIN standards, while the field above the M was reserved for the manufacturer's identification logo, initials, or code, and the field below was reserved for the material type marking.

Fig. 30: An example of a MPD marking, where "43" represents the manufacturer and "T" the pressing material classification.

the letters "M," "P," and "D" was developed in 1938 by and represented the *Staatliches Materialprüfungsamt*. A logo, abbreviation, or code representing the manufacturer was to appear in the field above the M, while the field below the M was reserved for the type of plastic and filler materials used in the item's manufacture.

5

The Manufacturing Process and Constituents

The vast majority of German *Kunststoff* products were made from phenol-formaldehyde or urea-formaldehyde resin and some sort of filler. The mixture of the resin with the filler is called a "pressing composition." Pressing compositions produced with phenolic resins were called *phenoplasts*, while those produced with urea resins were called *aminoplasts*.

Fig 31: Most processors and manufacturers of *Kunststoffe* during the war had their own research and development, and quality control laboratories, as depicted in this well equipped factory. *US NARA*

Fillers were an essential component of the pressing composition, and the qualities (and desired use or application) of the end product depended substantially on the type used. Fillers were available as powders, fibers, chips, flakes, shreds, or strips, and included such substances as sawdust, wood flour, wood chips, pulp, cotton flock, or talc.[8]

The synthetic resin and the filler material were heated and mixed together in certain proportions in kneading and mixing machines. At the time of this mixing, the desired color of pigment was added, and through various techniques, the effects of marbleizing, mottling, and spotting could be achieved.

Then, by adding certain chemicals, the rate of curing of the various mixtures could be controlled or manipulated. Some pressing compositions were slow curing and some faster. Most of the items made in wartime Germany appear to have been made from "quick press compositions."

After mixing, the composition could be crushed into a powder and used as was, with individual units measured and weighed for the intended end product. The use of tiny pellets or tablets, however, was more desirable and easier to work with. Powdered mixtures were

Fig. 32: A large tunnel dryer for phenolic molding powder used at the firm Bisterfeld & Stolting, Radevormwald. *From "German Plastics Practice", Office of the Quartermaster General, 1945.*

pressed and formed in pellet presses that reduced the overall volume of the mixture by some 1/3 to 1/10 of its original state.

Next, the tablets were pre-warmed and inserted into the mold(s). As excerpted from *Werkstoff Ratgeber*, these forms and:

"...pressing tools conclusively determine the quality of the pieces produced. They have to be manufactured with the greatest care and in most cases be polished inside to a high luster, and are therefore expensive. Naturally, one tries to get by with forms made of carbon steel or low alloy steel, which...[is ideal] with phenoplasts and with those aminoplasts which are made of urea...Instead of milling out the upper and lower part of the form with difficulty according to the copying process, in simple cases one presses the finish-worked pressing stamp ("priest") under high pressure into the die block of soft steel, whose recesses, worked only in advance, thereby already closely approach their final shape ("deep sinking").

Large work pieces require *single forms*. With forms for high-walled crates, containers or housings, the walls of the form are set up to be foldable. That facilitates withdrawal and the consideration of undercuts. Small parts which are required in large quantities are manufactured in *multiplex forms*, which often turn out 30 pieces, [even] up to 100 pieces at once. Filling frames are usually used for feeding the multiplex forms with pressing materials. Here the advantage of the tablets [as opposed to] the dusty powder shows itself quite clearly. If one places the pressing material pre-warmed to 70 to 80 degrees into the form, it can flow more uniformly and harden more quickly.

...most forms are built so that the little surplus of pressing material ["Flash"], which is as yet necessary so that one can be certain that the form is well filled, can leak and form a *ridge*. This can be made paper-thin and thus removed with extreme ease.

As soon as the filled form is closed, the pressing composition begins to become fluid under the effect of temperature and pressure; it then flows and fills out the form. Thereby the pressing stamp has taken its deepest position. According to the wall thickness, hardening in the form requires 1 to 5 minutes, sometimes even more."[9]

The process described above is now known as "Compression Molding," and is still in limited use today, though this method has largely been superseded by transfer and/or injection molding processes.

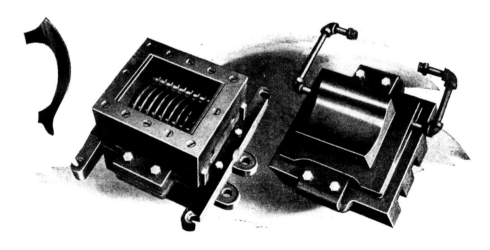

Fig. 33: A single form mold (above) and a multiplex mold (below) from the 1928 book *Bakelite Technic*.

Chapter 5: The Manufacturing Process and Constituents

Fig. 34: Another illustration from the same publication showing how bayonet joints are produced in a radio tube multiplex mold.

FIG. 60—Tilting head press

FIG. 62—Four-rod hot platen press

FIG. 61—Hot platen press

FIG. 63—Low pressure chilling press

Fig 35: Various presses used in phenolic processing, from *Bakelite Technic*.

Chapter 5: The Manufacturing Process and Constituents

Fig. 36: A monstrous 5000-ton Platens molding press, reportedly the largest in Europe. *From "German Plastics Practice", Office of the Quartermaster General, 1945.*

Fig. 37: An American investigator interviewing a German civilian worker in front of the same press, made by the Becker & van Hüllen firm. From "German Plastics Practice," Office of the Quartermaster General, 1945.

6
Types and Identification of Pressing Compositions

The 1943 *Werkstoff Ratgeber* described and outlined the approved pressing compositions, and provided a standardized listing and classification system for the types of materials then in use. Some types of pressing compositions were best suited as insulating material; others were suitable for food containers, while some types were useful in rigid applications, and others were made more flexible.

General purpose pressing compositions could be made with cheaper wood flour or pulp. End products that needed to be impact resistant were made with pressing compositions that contained long fiber fillers, such as cotton flock, cloth, or string. One example of an item requiring impact resistance was the butt stock of the MG34 machinegun. Allied appraisals of these reported, "The plastic stock and butt plate of the gun showed excellent strength and impact properties.... The gun stock and butt plate are phenolic resin, filled with macerated viscoserayon fabric and wood-flour." Heat resistant products and electrical insulators were made from compositions containing a large percentage of mineral fillers, such as talc or rock dust. A contemporary explanation of the differences follows:

"Type S is suited for general use. The types with textile fibers as filler material fulfill high mechanical demands, the types with inorganic filler materials have good heat stability. The types T3 and Z3 may be applied only with those parts with which the fiber binding is maintained substantially in its original arrangement while being pressed, thus preferably with flat pressed pieces. One can consider the types T3 and Z3 as layered materials."[10]

The synthetic resin, and the filler material before and during the early part of the war were heated and mixed together in about the same proportions in kneading and mixing machines.

In any event, the whole of the German plastics industry was expected to comply with the published standards and the manner of marking. As excerpted below:

"The warm-pressed synthetic resin pressing materials are normed according to DIN 7701. With respect to the classification of the pressing materials, the norm depends on the standardization…

Type	Group	Type of Filler Material and Remarks
11	Phenolic resin with inorganic filler	Rock dust (presumably talc)
12	Phenolic resin with inorganic filler	Asbestos fibers
M	Phenolic resin with inorganic filler	Asbestos cord
0	Phenolic resin with wood filler	(zero) Sawdust; composition for slow pressing
S	Phenolic resin with wood filler	Sawdust
T1	Phenolic resin with textile fiber filler	Short textile fibers
T2	Phenolic resin with textile fiber filler	Shredded textile fabric
T3	Phenolic resin with textile fiber filler	Textile fabric strips
Z1	Phenolic resin with cellulose filler	Short-fibered cellulose (flakes)
Z2	Phenolic resin with cellulose filler	Cellulose shreds (e.g., paper shreds)
Z3	Phenolic resin with cellulose filler	Layered cellulose (e.g., paper strips)

[the following are non-phenolic polymers, but were used to manufacture certain items in WWII on a smaller scale:]

A	Cellulose Acetate	[Celluloid; not used especially much in WWII]
K	Urea resin with organic filler	e.g., cellulose [as filler]."[11,12,13]

Chapter 6: Types and Identification of Pressing Compositions

However, as WWII dragged on and chemicals became scarcer, the *Kunststoffe* manufacturers had to improvise, frequently decreasing the amount of resins, adding more fillers, or substituting one material for another. The percentage of wood flour was:

"...constantly being increased during the war to the detriment of quality. Shockproof mixtures originally contained cloth, but toward the end of the war they used paper string which gave inferior quality."[14]

Of note, the author has observed a number of items from late 1944-45 that are marked with the standard MPD symbol with the material classification data omitted. The reason for these discrepancies is not exactly confirmed, but it is reasonable to conclude that the designation was left out to indicate that the pressing resins used were not up to established MPD and industry standards.

Fig. 38: Two *Zünderbüchsen* (artillery fuse containers) made by different firms in 1939, but using the same material, "S," or phenolic resin with fine sawdust used as a filler. Note the differences in hues that could be achieved. The container on the left is a smooth, glossy, deep brown, while the one on the right is pleasing brilliant orange-red.

Fig. 39: A binocular eyepiece protective cover made from "T2," or phenolic resin, with shredded textile fabric as filler.

Fig. 40: Close up of the same. Note the woven filler material mixed in with the phenolic resin.

Chapter 6: Types and Identification of Pressing Compositions

Fig. 41: The *Patronenausstoßer 15*, or "Cartridge Stripper 15," as used with the *Luftwaffe* MG15. This handy tool, used to strip cartridges from drums, was made from a very low percentage of phenolic resin and large wood or cellulose flakes.

Fig. 42: Same *Luftwaffe* cartridge stripper. Note the size of the cellulose flakes.

Fig. 43: The exterior of a 3.7 cm *Pak.* exercise or practice shell. While the base is made from steel, the body was formed from an interesting combination of multiple layers of phenolic resin-bonded, wrapped paper. *Courtesy Kenneth Huddle.*

7
Quality Control

Once completed, the finished product was inspected for quality against established standards. Each company was different. Larger firms had the finances and resources for more expensive testing machines. A common example in the era was called the "Schopper Dynostat" apparatus, used by many firms. Another testing machine, the "Schmidt-Bisterfeld Hot Needle Testing Apparatus," or *Wärmeprüfgerätmaschine, Schmidt-Bisterfeld*, was in limited use. It operated on the principle of measuring the penetration of a hot needle under a standard load into *Kunststoffe* against time.

Smaller firms had different testing procedures. A report from the era illustrated a much less high-tech testing sequence mandated for the inspection of finished *Hautentgiftungssalbe*, or "Skin decontamination salve" bottles:

"During production these bottles were subjected to careful inspection involving the following tests:

i. An air pressure test to see if the base was sealed on correctly.
ii. The bottle was filled with water and stored in ice to see whether it would withstand winter conditions.
iii. A 200 gram weight was allowed to fall 20 cm. on to the centre of the side of the bottle. The firm was allowed 2-3% breakage.
iv. As a static strength test a dead weight of 5 kg. was placed on the side of the bottle."[15]

8
Other Methods of Differentiating Various Materials

Not everything that appears to be is actually a wartime thermoset. The presence of the standardized pressing composition markings described above simplifies the matter. However, not all firms complied to the letter with the standards. So too, many items lack the MPD marking altogether, as some companies working exclusively for the military omitted them, and some of the firms in occupied countries making goods under German supervision did not feel the need to comply with "foreign" standards.

Luckily, there are a number of practical tests and cues to help determine from what a particular item was made, as follows:

Phenolics are the most cost effective of thermosets, and provide a good combination of strength, heat, chemical, and scratch resistance, and the end products can have excellent insulating qualities. Phenolic plastics of the era tend to have significantly more weight to

Fig. 44: A beautiful early wartime example of the Walther PPK pistol, with grip demonstrating the pleasing marbling effect that could be achieved by slightly mixing different hues into the phenolic pressing material.

them than modern plastics or other substances. The un-dyed material is naturally clear to cloudy, but this tends to darken or amber with age. Phenolics could not and cannot be produced in bright colors or "white." Fillers, such as wood pulp or chips, bits of fabric, etc. are frequently seen, as are mottling and marbling techniques. When placed under hot tap water and rubbed with a finger, a phenolic frequently gives off a faint fish-like (carbolic acid) odor.

Urea-based polymer plastics are quite similar, and even superior in many regards to the phenolics, but were more expensive to produce. The vast majority of the country's production of urea resins was used to produce adhesives, with only a small fraction earmarked for use as molding compounds for end-products before and during the war.

Urea-formaldehydes were very durable, naturally clear, and could be formed with a number of fillers and dyed any color, to include brighter hues. Unlike the phenolics, though, the urea-based polymer plastics do not tend to change color with age, nor do they impart taste or odor to food—marked commercial advantages over the phenolics. It was quite brittle, though, and tended to crack or chip easily.

An offshoot of urea based plastics was Melamine. Melamine was and still is noted for having better water resistance, hardness, and stability than the earlier urea-formaldehydes. In its un-dyed state Melamine is colorless, and could be combined with just about any dye or filler (to include pastels), being generally color-fast over time. On the downside, Melamine was even more expensive to produce, and during the war was used mostly in the manufacture of tableware and food containers. Both urea and melamine plastics are

Fig. 45: Various large and small bore exercise cartridges made from black and red tinted *Kunststoff*. Courtesy Kenneth Huddle.

amino based; therefore, when placed under hot water or rubbed, they may tend to smell like ammonia.

Polystyrene was a fairly "new" plastic polymer by the eve of WWII, made from erethylene and benzene. This versatile plastic has bending properties similar to rubber, and can be made in bright colors, such as red or orange. Noted first by Eduard Simon in 1839, it was not until a German chemist, Hermann Staudinger, began working on it in earnest that its potential was appreciated. He published his findings in 1922,[16] and in 1930 the scientists at BASF (Badische Anilin & Soda-Fabrik, a subsidiary of I.G. Farben) developed a way to commercially manufacture polystyrene.

The author has not noted any particular scent associated with polystyrene end-products during the era, and the material does not appear to have been used to a great extent during the war. This was likely due to critical shortages of its chemical precursors, particularly benzene. Polystyrene was used to some extent to make anti-chemical agent ointment bottles and other containers, grip plates, and presumably formed the plastic portion of most exercise cartridges of the era.

Casein is a polymer plastic derived from milk protein and formaldehyde, which was traditionally used to simulate bone, horn, or ivory. Only a few manufacturers manufactured casein products in the 1930s, and this material was not utilized much during the war:

"Owing to the shortage of Rennet Casein, the manufacture of casein plastics in Germany had ceased in 1941, excepting only the Galalith Co. [Internationale Galalith-Gesellschaft or I.G.G. located in Hamburg] who continued to manufacture on a reduced scale until 1944…for the same reason, most of the machinery was idle for four years…."[17]

When encountered, casein products are white or off-white, and it is said that when rubbed they may have a slight odor similar to scorched milk.

Celluloid, on the other hand, was still being used to a moderate extent prior to and after the war began. *Zelluloid*, as it was known in German, was usually plain, clear to opaque in appearance, and usually did not contain visible fillers, but hues were blended and could be marbled. A slightly softer and more flexible (in thin sections) material than the phenolics, when warmed Celluloid may smell like vinegar, camphor, or moth balls.

Hardened rubber is simply rubber that has been over-vulcanized, to the extent that it loses most of its elasticity. Almost always produced in a matte or semi-gloss black, this material can be mistaken for phenolic polymers that have been mixed with black pigment. Hardened rubber was frequently used for pistol grip plates. It displayed some bending properties that wartime phenolics usually do not. It also tends to be heavier than the phenolics, and used in smaller sized applications. When rubbed it may give off a slight sulfur or "burnt match" odor.

Lastly, "Durofol" is a brand name of an interesting material produced during the war that is well known to many small arms collectors. In appearance, Durofol looks like Bakelite or a phenol-formaldehyde plastic, and many collectors call items made of it "Bakelite." This is, however, technically incorrect. Firstly, both names are proprietary and unrelated.

The material and the trademark have been attributed to the firm of "Durofol K.G., O. Brangs & Co." in past publications. However, this could not be absolutely confirmed in any contemporary publication. No entry related to the firm "Brangs" could be found in any wartime or post-war "code" books or listings, either. The Durofol trademark was originally registered on March 17, 1942, and held at least since 1951 by Hans-Joachim Gerling of Solingen. "Durofol Preßwerk, Solingen" filed a number of patents in the 1950s, some of which use or refer to *Preßschichtholz* (literally "pressed laminated wood") and/or *Kunstharzpreßholz* ("synthetic resin pressed wood). "Brangs & Heinrich" is an old established, prominent industrial packaging firm with some current laminated wood products headquartered in Solingen as well.

All of these connections to Solingen are tantalizing, and are likely to be much more than coincidence. But neither Brangs nor Durofol were listed as a plastics manufacturer/plastics material in any of the wartime plastics documents or lists that the author reviewed. The reason for this is clear: Durofol is not a plastic, but rather, a "plywood" or laminate made up of long slices of, usually beech, pressure formed and bonded together with a synthetic resin.

Upon close inspection of items marked with the Durofol trademark, one can see long beech veneers, or "plies," that make up the bulk of the material. This, and the complicated, curvaceous shapes of the products marked Durofol, add further credence to the process of pressure bonding and forming. Likewise, the suffix "fol," in German chemical jargon of the time, was reserved for laminated wood, and other companies are confirmed to have made similar products with trade names such as Ligno*fol*, Dehe*fol*, etc. Even Albert Speer himself wrote in his book *Infiltration* that Durofol was "…compressed wood."[18]

Government investigators/authors of the 1946 *German Plastics Practice*, under the section "laminates," reported:

"In laminated wood Durofol was produced by impregnating beech veneer with 30% of [a] water-soluble resin. This was made by [Heinrich] Ritter in Esslingen, was used for bayonet scabbards, and as might be expected from the type of resin used was somewhat deficient in impact strength."[19]

Chapter 8: Other Methods of Differentiating Various Materials

Ritter is known to have sold wooden covered canteens marked in cursive "Ritter," followed by the initials "H.R.E. [Heinrich Ritter, Esslingen]." Ritter may have made Durofol, but since the firm was (and still is) a longtime aluminum and light metal ware company, it may have simply purchased it for use in Esslingen. Unmarked pressed wood scabbards are also known to have been used with the SG42 bayonet, and of course both G43 rifle hand guards and P38 pistol grip plates are frequently encountered marked with the Durofol trademark. As such, further corporate, patent, and trademark research is clearly warranted.

Unfortunately, the 1946 report does not go on to describe the actual production process. End products marked Durofol show no signs of milling or sanding, indicating that the finished state existed for the most part when an item left the forming dies. It is assumed that the adhesive used to hold the veneers together would be forced to the exterior and interior surfaces of the forming die under heat and pressure, thus explaining the "Bakelite-ish" external appearance of finished products. The exact context of "water-soluble resin" is also not clear, as it could be taken to mean that the resin was water-soluble only prior to the

Fig. 46: Two Walther-designed *Gewehr 41* rifles (top, second from top). The G41 was issued with both black and brown *Kunststoffe* handguards, all of which were made by Presswerk A.G., Essen. Below the G41s are two later G43s, also designed by Walther. The rifle second from the bottom was issued with a traditional laminated wood handguard, while the Berlin-Lübecker Maschinenfabrik manufactured rifle below is equipped with a Durofol handguard (bottom). *David W. Davis collection*

Fig. 47: Two hand guards for the *Gewehr 43* rifle. Hand guards machined from laminated beech plywood (above) were susceptible to longitudinal cracking under recoil, and were labor intensive to produce. Hand guards made from Durofol (below) took less time to produce, and were nominally stronger (though they too tend to crack/break at key stress areas).

Fig. 48: Internal comparison of the same hand guards.

Fig. 49: A close-up of the markings on the inside of a phenolic resin bonded compressed wood hand guard. Apparently, the sole manufacturer of these was Durofol K.G., O. Brangs & Co. Hand guards made from Durofol are marked on the inside with a "1," "2," "3," etc. representing form position numbers. They appear to be exclusive to G/K43 rifles produced by Berlin-Lübecker Maschinenfabrik.

Chapter 8: Other Methods of Differentiating Various Materials

heating and pressing process. Certainly, the author has not observed any water or moisture degradation, separation of veneers, etc. in any wartime pressure formed laminated wood items.[20]

In any case, Durofol was initially met with fabulous enthusiasm when it was introduced during the war. The Nazi hierarchy even anticipated the phenomenal new material replacing steel and ingot iron. Indeed, T-girders, I-beams, and even the bodies of automobiles were all expected to eventually made from it. But these fantastic expectations did not come to fruition:

"On June 25, a disillusioning Jüttner reported that this Durofol was artificially compressed [compression molded, laminated] wood.[21]...lack of elasticity limits the applicability of the material so that it cannot be used for manufacturing the bodies of automotive vehicles. It would be suitable for smaller parts...gears, doors, door handles, window frames and the like...basically a substitute for nonferrous metals and for aluminum. At the moment only for smaller objects up to around 500 millimeters...."[22]

Fig. 51: Wartime LP42 flare pistol made by C. & W. Meinel-Scholer, Metallwaren, Klingenthal, with unmarked resin-bonded, pressed wooden grips.

Fig. 50: German *Tropen Feldflasche 31* canteen made in 1943, colloquially known as the "tropical Bakelite" type. However, the material used to form the exterior body of these examples does not appear to be a phenolic polymer, but rather, a form of pressure-bonded wood, similar to Durofol. This example is marked "H.R.E. 43."

9

Examples of Manufacturers

Hundreds of manufacturers were involved in producing the raw materials, creating the resins, or processing polymer plastics into end products in wartime Germany, as well as a number of companies in the occupied territories. A discussion of each would of course not be practical. Generally speaking, the larger chemical concerns made the bulk of the country's thermoset resins, and sold these to smaller companies to mold into finished goods, but most had their own molding facilities as well.

The August Nowack AG firm in Bautzen was a typical concern, run by the Kopp brothers. It had a production capacity of around 400 tons per month of phenolic pressing materials, and also made a limited amount of varnishes.[23]

Hermann Römmler AG, Spremberg, on the other hand bought most all of its pressing material from larger companies, and processed about 300-400 tons of phenol-formaldehydes in their molding shop into finished products.[24]

Dr. Kurt Albert G.m.b.H., Chemische Fabriken, located in Amöneburg, near Wiesbaden-Biebrich, had a capacity of about 500 tons per month, 300 of which was general purpose material mixed with wood flour, and the rest being ammonia-free or impregnated paper sheet.[25]

Internationale Galalith-Gesellschaft was a fairly large concern, located in Hamburg-Harburg, which prior to the war had manufactured products from casein, and supplied materials to smaller firms for the same. When the supply of domestic Rennet Casein dried up, the firm shifted much of its capacity to phenolic resin and molding powder processing. IGG also made some finished products using phenolic resins via both injection and compression molding techniques. During the war the firm employed some 600 full time workers, and had a number of mills, mixers, extruders, platen presses, injection as well as compression molding machines, and some 25 blocking presses.[26]

Dr. Fritz Raschig Chemische Fabrik, GmbH, located in Ludwigshafen, was a much larger firm involved in all sorts of petro-chemical and *Kunststoffe* production. Described as a "pioneer in the field of phenol-formaldehyde resins," Raschig alone produced some 4.5 tons of phenol daily, 300 tons per month of phenolic molding powders, and another 30 tons per month of its phenolic "cast" resins. It also had another factory located in Leipzig, and between the two also produced around 3000 tons of phenolic adhesives annually.

Dynamit AG, part of the I.G. Farbenindustrie group, was perhaps the biggest company involved in *Kunststoffe* production before and during the war. The concern had a controlling interest in some forty factories, and plants extended throughout Germany, which were directed from Troisdorf. At Troisdorf itself, the company employed some 9,000-10,000 workers, of which some 2,000 were forced or relocated laborers. Of this number, about 4,500 personnel worked in the plastics division, "Venditor"; 1,000 of them engaged in machine design and engineering, essential to the operation of the facility.[27] Only about half of the company's effort was devoted to plastics manufacture, with the remainder involved in explosives and propellants research, development, and production. Buildings were generally small and spread out, but well staffed, equipped, and tooled. Dynamit produced about 700 tons per month of Celluloid material, 1000 tons per month of phenolic molding powders, and 500 tons of urea-formaldehydes. Most of its wartime *Kunststoffe* production was concentrated on phenol and urea-formaldehyde based products and adhesives, but the firm is also known to have produced significant amounts of Melamine, and supplied a number of other firms with tooling, machinery, and molds.[28]

"Phenol-formaldehyde resins of "Novolak" and "Resol" type [trademarked material] constituted another of the factory's manufactures, conventional compression mouldings of multifarious character being made. The factory was stated to have prepared large numbers of field telephones for the German army, cups, combs, handles, etc.; mouldings examined being of good quality and workmanship."[29]

During 1944 and 1945, the firm's entire production went toward the war effort.[30]

10
Manufacturers' Logos, Markings, and Codes

In the 1920-1930s many of the larger *Kunststoffe* firms identified the products they made by molding either their company logos, or the initials of the company name into the finished products. Between 1933-and the early 1940s, company name abbreviations are often seen molded inside the MPD marking, but have also been observed adjacent to the marking, or even alone on certain products. A short list of these abbreviations and the manufacturers they represent, as excerpted from the trade journal *Kunststoffe,* is listed in Appendix III. Additional abbreviations observed, confirmed, and catalogued by the author during the course of study and research are also listed.

The National Socialist regime made the decision in 1933 to secretly rearm Germany in violation of the Treaty of Versailles. Manufacturers of arms and ammunition were soon instructed to adopt and mark their products destined for the military and para-military organizations with number codes, instead of their traditional company trademarks and logos. This was an attempt to disguise the point of origin of these products from the watchful eye of potential enemies trying to ascertain Germany's level of compliance with the treaty.

The Simson firm in Suhl was the only company authorized by the Treaty of Versailles to make small arms for the German military. When the Nazis ordered clandestine production to commence at other factories, they followed a general policy of secrecy. Cloaking the actual location by grouping them all under Simson's "S" designation, they also added an additional number suffix to designate the true point of origin. If a potential adversary saw a rifle marked "S/42" dated from 1934 onward, for instance, it would be explained that the rifle had come from Simson, as obviously indicated by the "S." "S/42," however, did not really represent Simson, but rather, is widely known to have been used by the Mauser firm. S/27, S/147, S/237, and S/243 are other examples of well known rifle makers. The letter "P" is likewise known to have represented Polte Patronenfabrik of Magdeburg, the only

Chapter 10: Manufacturers' Logos, Markings, and Codes

campany in Germany authorized under the treaty to produce ammunintion, or *Patronen*. Under this same secretive policy, other companies later tasked with cartridge production were also assigned the umbrella letter "P," but this was followed by their own numeric designation. Later, the letter prefixes were dropped from these clandestine numbers. Dropping the "S," small arms makers used numeric designators, such as "27," "147," and "237" for a short time in the late 1930s.

About three dozen firms involved in *Kunststoffe* production appear to have been associated with these secretive number designators during the period. But, while they may have been assigned a number, these designations were only very rarely actually molded into end-products. All of these companies were well established manufacturers. The reason they were given numbers in the first place probably has much to do with the types of forbidden ordnance products, munitions, or communications related equipment they were involved in producing, rather than plastics manufacture per se. Additionally, prefixes such as "S," "P," "Rhs," et al do not appear to have been used by any other member of the *Kunststoffe* industry proper during this same pre-war National Socialist timeframe. The plastics industry after all was not mentioned in or governed by the terms of the Versailles Treaty, and at that point didn't appear to be in any real need of shielding the identity or point of origin from the outside world.

Fig. 52: A fairly rare example of a plastics manufacturer's use of the secretive number coding system designed to cloak armaments and ordnance production from the Inter-Allied Control Commission. In this case the number "950" code represented the H. Römmler AG, Preßstoffwerk of Spremberg. *Courtesy Kenneth Huddle.*

Instead, German plastics companies almost always simply continued to use their initials, abbreviations, proprietary trademarks, or logos to represent their firms. Product "branding" as a concept was just as important then as it is today, and it was in a firm's best interest, from a marketing stance, to proudly mold its brand, trademark, or logo into products destined for both the military and domestic commercial and export markets.

Fig. 53: A wartime-era individual soldier's flashlight assembled and marketed by Zeiler. The housing, however, was made by H. Römmler Aktiengesellschaft, Spremberg, and marked additionally with that company's logo, "HRS."

Fig. 54: Initials molded into a *Kunststoff* container indicating manufacture by Fr. Möller, Brackwede i. Westf. *Courtesy Kenneth Huddle*

Chapter 10: Manufacturers' Logos, Markings, and Codes   65

Fig. 55: A phenolic tent peg made in 1941, and marked with the manufacturer's initials "GS," representing Gebr. Spindler KG, Köppelsdorf i. Thür.

Fig. 56: HRS trademark representing the large H. Römmler Aktiengesellschaft of Spremberg firm.

Fig. 57: Initials molded into another container that represented Kunstharz-Presserei Schwaben, Ingenieur Otto Single, Plochingen i. Württ. *Courtesy Kenneth Huddle*

Fig. 58: "Cewe" trademark, as molded into this flare pistol grip plate made by Carl Walther, Zella-Mehlis. *Courtesy Kenneth Huddle*

Fig. 59: "Ph" in a shield, a trademark of Phenoplast, Bischoff & Co., Kom.-Ges., Eberswalde. *Courtesy Kenneth Huddle*

Fig. 60: Frequently encountered Presswerk A.G. trademark. *Courtesy Kenneth Huddle*

Chapter 10: Manufacturers' Logos, Markings, and Codes

Fig. 61: A small ordnance shipping cap bearing the stylized logo "CAW," a trademark of Caspar. Arnold Winkhaus firm, located in Carthausen.

Fig. 62 (above): A small black phenolic grease container with screw off lid, used to store machine gun lubricant.

Fig. 63 (left): A currently unidentified "HK" maker's logo as seen on the same grease container.

Kunststoffe

Fig. 64 (above): Inside view of a pair of SG 84/98 bayonet grip plates. The manufacturer's stylized "G" trademark is not confirmed. Fig. 65 (below): The logo of the firm BEBRIT-Preßstoffwerke G.m.b.H., Bebra (Hesse) and C & F Schlothauer GmbH, Ruhla (Thür.) as molded into an igniter container. *Courtesy Kenneth Huddle*

Trademarks and logos have been observed molded within the confines of the MPD marking, adjacent to it or alone. Unfortunately, no complete contemporary list of *Kunststoffe* manufacturers' trademarks appears to exist, or at least was not discovered during the author's research. Some companies owned multiple logos and trademarks. Other firms that assembled products that contained only a percentage of *Kunststoff* components subcontracted the polymer plastic portion to other firms, and the end items bear logos and trademarks of the main contractor, and frequently the MPD marking of the plastics manufacturer, as well. A list of logos and trademarks with depictions of the same as observed by the author on pre- and wartime German examples is provided as Appendix II; though by no means is this list complete, given the multitude of companies involved.

Two-Digit Numbers Representing Kunststoffe Manufacturers

Non-standardized initials and trademarked logos were problematic; often variable, difficult to index or catalog, and some could not be read or reproduced small enough to fit into the tiny MPD marking. Some members of the *Technische Vereinigung der Hersteller und Verarbeiter typisierter Kunststoff-Formmassen e.V.* began lobbying for a better, more uniform system, until eventually a consensus was reached:

"The manufacturers of the pressing materials have agreed upon an identification of their products through trade numbers."[31]

A number of sources contend that these two-digit *Kunststoffe* trade numbers originated in 1938, but this is not correct. Use of the numbers in conjunction with the MPD marking have been observed on products dated as early as 1937, and the author speculates that the idea of industrial trade numbers may date as early as 1936.

Other sources and collectors are convinced that the assignment of these numbers by the MPD corresponded, or was in conjunction with the previously discussed effort to shield arms and ammunition makers from the watchful eyes of potential enemies by assigning secretive number designators to those firms. This does not appear to be true. Firstly, the use of numbers by the small arms and ammunition manufacturers appears to predate the *Kunststoffe* number designators by at least three years. Additionally, observations confirm that some small arms manufacturers used the same number as some of the plastics manufacturers.[32] Lastly, of the manufacturers confirmed to have been assigned both sorts of number codes, no "matching" was observed. Just one of the many examples, Isola-Werke A.G., Birkesdorf -Düren (Rhld.), was assigned the MPD designation "40," but was also assigned the number designation "651" to shield its activities from foreign governments.

Thus, the author's conclusion is that the *Kunststoffe* number assignments were instituted by the trade association through the MPD, coincidentally and independently from the other more strategic effort to maintain military-industrial secrecy.

In any case, the first listing of firms and designations discovered by the author as tabulated and consolidated by the *Staatliches Materialprüfungsamt* was published in December 1938. Once assigned and published, each *Kunststoffe* manufacturer was presumably expected to comply, omit their logos or initials, and use the designation assigned. Like any issue involving the corrupt National Socialist era, compliance with this standard was spotty. Some makers continued to mold in their logos and/or initials, either within the MPD marking, or on other areas of the item up to the end of WWII. Others molded in their company logos on the piece *and* used their numeric or alpha-numeric designations inside the MPD marking on the same item as well. As the war began and then dragged on, however, these practices became thought of more and more as strategically problematic and became less common, though neither practice appears to have ceased entirely.

The complete list of *Kunststoffe* manufacturers in alphabetical order with their assigned MPD codes is provided as a ready reference as Appendix IV. Appendix V lists only the manufacturers that were assigned two-digit numbers, extracted from and transcribed exactly as they were originally written in the wartime publications, and subsequently arranged by the author in numerical order for ease of reference.

Of note, the available lists of firms assigned these numbers begin at the number "21." The reason that it does not start at "1" and no firms were assigned numbers 1-20 is not

Fig. 66: A *Zünderbüchse* container made in 1939 by Süddeutsche Isolatoren-Werke G.m.b.H, Freiburg, which chose to use both the company logo and its "25" numeric designator as assigned by the MPD. The "S" in the bottom field of the MPD marking indicates that the container was made from phenolic resin, with sawdust used as a filler.

Chapter 10: Manufacturers' Logos, Markings, and Codes   71

known. The author speculates that since the material type classifications "1," "2," "3," "4," "6," "7," "8," "11," "12," and "0" were already in use, and published via the *Staatliches Materialprüfungsamt*, that the agency did not assign numbers 1-20 to represent any firms so as to avoid confusion. It is also possible that someone in the MPD might have felt that these and certain other unassigned numbers had the possibility of being confused with other non-plastics companies, or perhaps mistaken for two-digit date markings (such as numbers 41, 42, 44). Likewise, it is entirely possible that the unlisted numbers were previously assigned to firms that had since been acquired by larger companies, or those that perhaps had simply ceased operations.

Also, although lists from 1938, 1939, and 1940 are known and referenced here, the author did not discover any subsequent listing (i.e. 1941-1945) during the research for this work. It is assumed that they exist, and may list additional firms, codes, designators, and changes in company ownership or location.

Fig. 67: The "Venditor" trademark used by Dynamit-Actien-Gesellschaft vormals Alfred Nobel & Co., Abteilung Celluloid-und-Kunstoff-Fabrik, Werk Troisdorf incorporated into the mold used to make this 1938 dated black phenolic binocular case.

Fig. 68: A *Zdlg.36* container made in 1941 by Dynamit-Actien-Gesellschaft, formerly Alfred Nobel & Co., Abteilung Celluloid-und-Kunststoff-Fabrik, Werk Troisdorf. Note that the firm incorporated another of its company trademarks, as well as its MPD mandated number designation "43," into the mold.

Letter-Number and Number-Letter Combinations Representing Kunststoffe Manufacturers

Two digit numbers easily fit into the small MPD marking, but once these were all assigned (or purposely not used) up to number 99, something else had to be done. The MPD must have identified this problem right from the start, as the makers and processors of German wartime polymer plastics eventually numbered in the hundreds. Adding another digit would have worked, but since a three digit number would not easily fit into the proper field of the already cramped MPD marking, two other code variations were approved.

It appears that two-number designations were originally assigned to the larger and more well established firms, and others were assigned either a letter-number or number-letter code. Some firms have more than one code for the same location, or different codes for other locations or subsidiaries. Appendix VI lists only the manufacturers that were assigned letter-number designators extracted from and transcribed exactly as they were originally written in the wartime publications, and subsequently arranged by the author in alphabetical order for reference. Likewise, Appendix VII is also provided in numerical

Fig. 69: A pair of brown K98 grip plates. Note the MPD marking "53" indicating Ernst Backhaus & Co., Kierspe, as the manufacturer. Z3 represents the pressing material used, Phenolic resin with layered cellulose (e.g., paper strips). The numbers 15 and 16 are mold position numbers.

Chapter 10: Manufacturers' Logos, Markings, and Codes

Fig. 70: Black canteen cap made by Gebr. Spindler KG, Köppelsdorf, as indicated by the alphanumeric designation "F7."

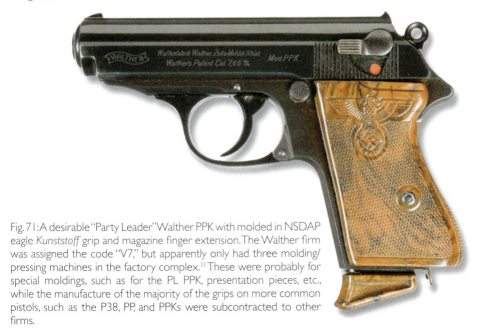

Fig. 71: A desirable "Party Leader" Walther PPK with molded in NSDAP eagle *Kunststoff* grip and magazine finger extension. The Walther firm was assigned the code "V7," but apparently only had three molding/pressing machines in the factory complex.[33] These were probably for special moldings, such as for the PL PPK, presentation pieces, etc., while the manufacture of the majority of the grips on more common pistols, such as the P38, PP, and PPKs were subcontracted to other firms.

Fig. 72: Inside of a fuse container, wherein the manufacturer, Emil Adolff, Abt. Kunstharz-Presswerk, Reutlingen, utilized its trademark, as well as its assigned MPD designation "H2." *Courtesy Kenneth Huddle*

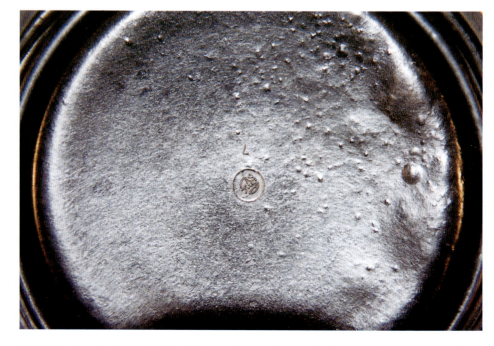

Fig. 73: Markings inside a typical *Fettbüchse*, or fat container, made from urea-formaldehyde by Gebr. Dahlhaus, Schalksmühle.

Chapter 10: Manufacturers' Logos, Markings, and Codes

Fig. 74: MPD marking on a black 6x30 binocular protective case made by C. Pose, Wehrausrüstungen, Berlin (0F), and made from phenolic resin with shredded textile fabric (T2) as filler.

Fig. 75: The reverse side of a typical *Heer* marching compass, with at least the phenolic body made by F.G. Zieger of Roßwein, as indicated by the designation "0W."

Fig. 76: A *Luftwaffe* electrical connector made by a company (thought to be Heinrich List Elektro-Meßgerätebau, Berlin-Steglitz) using the "List" trademark.

Fig. 77: Same connector. Note the trademark again, and the designation "4Z," a designation beyond those listed in the 1940 issue of *Kunststoffe*.

order, listing those firms assigned number-letter combinations. As with the two-number designations, there are unexplained gaps in these sequences as well, which may represent designations that may have been confused with other materials or firms that had since been acquired by larger companies, or those that perhaps had simply ceased operations.

Of note, the author has observed number-letter combinations beyond those listed in the 1940 listing as published in the journal *Kunststoffe*. This lends credence to the presumption that subsequent listings probably exist, and that companies that commenced operations after 1940 were likely assigned these currently unconfirmed designations.

Military Letter Codes

German small arms and ammunition manufacturers were ordered by law on July 8, 1939, to substitute an assigned number in place of their logos and trademarks, but as has already been discussed, most of the prominent firms had already been complying since the early days of the *Third Reich*.[34]

On September 1, 1939, Hitler ordered the *Wehrmacht* to invade Poland. Western Europe was next, and although the nation was at war and the *Kunststoffe* manufacturers shifted more and more of their production from consumer goods to military items, security within the industry was still pretty *laissez faire*. Goods were marked as they had been before, with at least the stylized MPD marking, as well as the assigned numeric or alphanumeric codes to identify each item's manufacturer. Many firms, perhaps a majority, even still continued to use initials or their company trademarks and logos adjacent to or within the MPD marking in this early period of "phoney" or "easy" war.

Additionally, it must have been assumed by many within the *Kunststoffe* industry that the prominent makers and locations were already widely known. Indeed, the first of the MPD lists was published in 1938, another in 1939, and still another in 1940 in the trade journal *Kunststoffe*, which was widely disseminated within the industry to both domestic and foreign subscribers, and thus were certainly already available to Allied intelligence officers prior to, and even into the first year of the war.

The British first began small scale leaflet dropping and actual sporadic bombing of German targets in the autumn of 1939. RAF aircrews then conducted their first large scale bombing of Berlin (in retaliation for an accidental bombing of London by disoriented *Luftwaffe* members) in August 1940, and by the end of the year had dropped some 14,000 tons of ordnance over the country.

These raids created quite a concern among the Nazi hierarchy, and many industrialists as well. It was apparent that the Allies were not just bombing military complexes, key infrastructure and traditional armaments manufacturers, but any industrial target of opportunity that might be vital to the German war effort. Clothiers, leather goods makers,

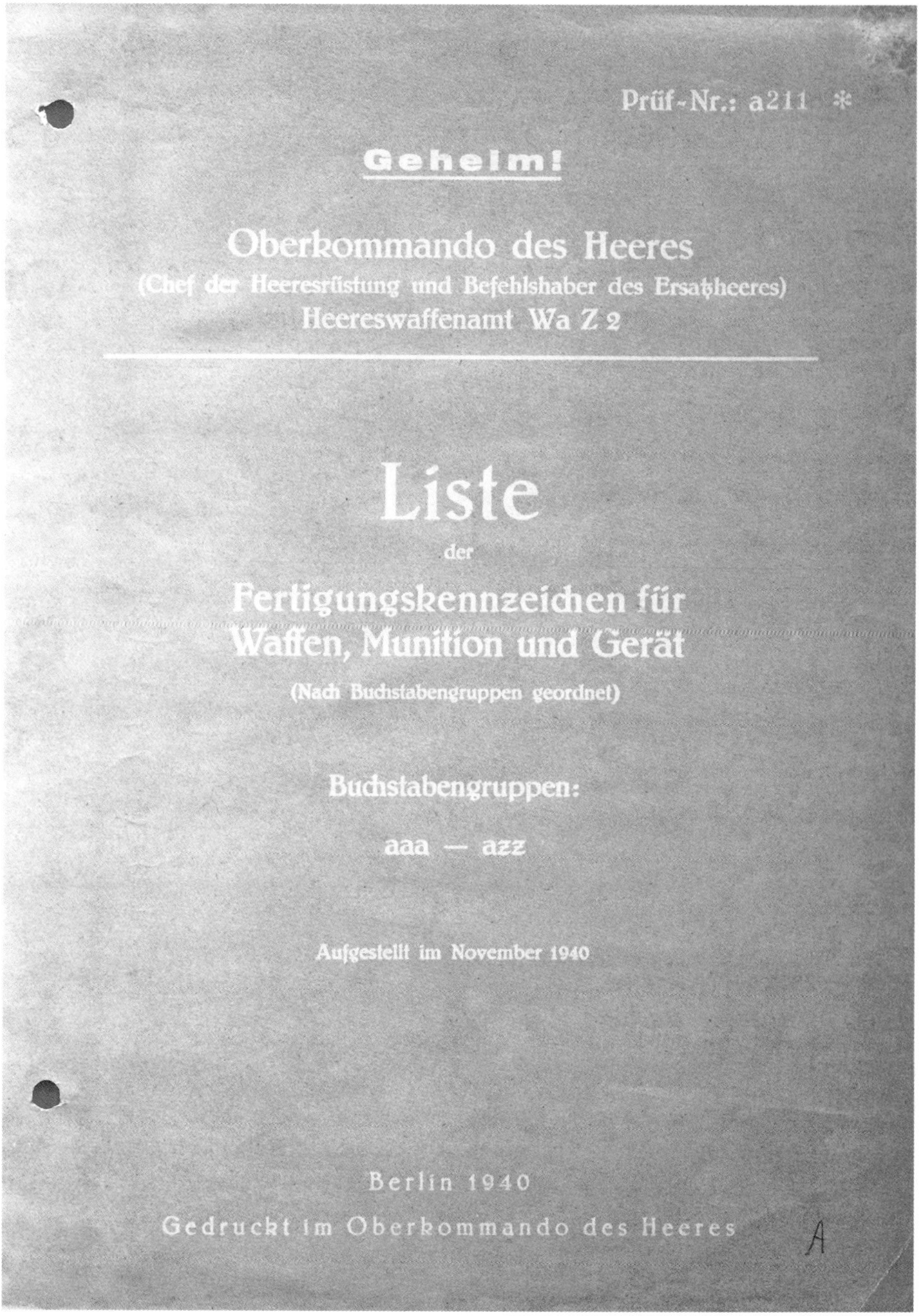

Prüf-Nr.: a211 *

Geheim!

Oberkommando des Heeres
(Chef der Heeresrüstung und Befehlshaber des Ersatzheeres)
Heereswaffenamt Wa Z 2

Liste
der
Fertigungskennzeichen für
Waffen, Munition und Gerät
(Nach Buchstabengruppen geordnet)

Buchstabengruppen:
aaa — azz

Aufgestellt im November 1940

Berlin 1940
Gedruckt im Oberkommando des Heeres

A

Fig. 78: The cover of the then highly classified first list of military letter codes, or *List of Production Identifiers for Weapons, Munitions and Equipment (Arranged According to Letter Groups), Letter Groups aaa-azz*, published by the *Oberkommando des Heeres* in 1940. *Courtesy US Army Ordnance Museum*

Chapter 10: Manufacturers' Logos, Markings, and Codes

foundries, precision optical manufacturers, and a myriad of others that had not previously needed shielding from the defunct Inter-allied Military Control Commission were at this point susceptible to identification and attack.

Kunststoffe firms were particularly vulnerable, and phenolics were becoming increasingly important to the German war effort, as components of all sorts of martial items from fuses to munitions and communications equipment. A number of these firms were also involved in petro-chemical, propellant, and explosives production, as well. Many tended to be clustered in key areas that would not have been difficult for Allied intelligence and targeting officers to identify through examination of a trademarked or MPD coded item. Thüringen (Thuringia), and to a greater extent Nordrhein-Westfalen (Westphalia), and the town of Kierspe in particular, were centrally important to the production of phenolics, and the processing of plastics dominated the local economy. In fact, on the eve of WWII about half of the German factories and manufacturers of phenolic goods were located in or near the area. Kierspe alone contained thirty-six companies involved in processing phenolics, and its plastics industry was equipped with some 400-500 phenolic pressing machines.

A new military letter coding system—vastly broader than the older numerical method—was developed by the nation's armaments hierarchy in an attempt to better cloak the identity and location of key manufacturers. These military letter codes were implemented in November 1940, beginning with the group "aaa-aaz," and would eventually reach "ozz" and beyond. *Kunststoffe* manufacturers, whose products were becoming more strategic than ever, were also assigned and presumably ordered to use these letter codes as well.

Interestingly, it appears that the new national letter coding system took precedence over the previously developed MPD numeric and alphanumeric system, though it is not known if it ever was intended to completely replace the old method of marking altogether. Compliance with the new system varied considerably, especially during the early years of the war. Many of the manufacturers adapted by placing their new one, two, or three letter codes inside the traditional MPD marking in place of their previously assigned MPD numerals or alphanumeric codes. Others continued to use the same MPD designations in the MPD marking, and placed the new letter codes adjacent to it. The most frequently seen method observed, particularly as the war dragged on, was to leave the manufacturer's field in the MPD marking blank while retaining the material classification code intact, and then place the new letter code somewhere nearby. Still others, though, went right on using their older initials or trademarks as they had done before.

A partial list of these confirmed two and three letter codes assigned and used by plastics manufacturers has been catalogued by the author, and included as Appendix VIII. Of note, a number of firms in annexed or occupied countries[35] that made *Kunststoffe* under German supervision were also ordered to use military letter codes, but do not appear to have been assigned an MPD designation by the *Staatliches Materialprüfungsamt* in Berlin-Dahlem.

Fig. 79: 1942-dated fuse container made by Hans Büllmann Werke für Elektrotechnik und Feinmechanik, Gablonz-Schlag (Reichsgau Sudetenland), as indicated by the military letter code "amh."

Fig. 80: The marking "ehj" as seen on a *Patronenausstoßer 15* made by Gebrüder Spindler of Köppelsdorf.

Chapter 10: Manufacturers' Logos, Markings, and Codes

Fig. 81: A MP40 grip plate marked "ayf" for Erfurter Maschinenfabrik Berthold Geipel GmbH (ERMA), but bearing the MPD code of AEG (Allgemeine Elektricitäts-Gesellschaft), Fabriken Hennigsdorf, the *Kunststoff* subcontractor.

Fig. 82: Inside of a fuse container made by Venditor Kunststoff-Verkaufges m.b.H., Troisdorf, as indicated by the military letter code "boa." *Courtesy Kenneth Huddle*

A concerted effort was initiated in the middle of the war to catalogue and reference manufacturers of military products under an even more standardized numbering system. Each individual contractor or manufacturer was assigned a *Reichsbetriebsnummer*, or roughly "National Contractor Number," which was specific to the firm or a factory. Apparently, the numbers were originally intended for written correspondence, contracting, requisitioning, and ordering, and many companies had these numbers pre-printed in the heading of their official stationary or invoices. However, as the war dragged on into 1943 a great many firms, especially those involved in cutlery, leather goods, clothing, and others, started stamping their products with their "R.B.Nr." in place of a company logo or previously assigned military letter code.

In contrast, this practice does not appear to have occurred within the *Kunststoff* industry at all; those firms kept their traditional trademarks, MPD designators, or military letter codes.

The Heereswaffenamt and German Industry

The MPD marking was fine for pre-war and commercial goods, but items destined for the German military were another matter. If it was going to be procured and issued to the troops, the *Staatliches Materialprüfungsamt*'s codes, DIN standards, and product requirements were in many cases of little or no consequence. German ordnance and material officials had their own specifications, inspection, and acceptance standards, as well as methods of marking military items.

Beginning in 1935 German small arms, ordnance, and equipment requirements were issued by a branch of the Ministry of War. In 1938 the War Ministry was abolished by Hitler, and replaced by the German High Command (*OKW* or *Oberkommando der Wehrmacht*). Under the new structure, a branch of the *OKW* became responsible for weapons and equipment, known as the Army Weapons Office (*Heereswaffenamt* or *HWaA*).[36]

The *HWaA* was itself a branch of the Army High Command (*OKH* or *Oberkommando des Heer*). The *Luftwaffe* and *Kriegsmarine* both had similar agencies. Generally, once requirements and/or contracts were issued to German industry through one of the three agencies, production of the item was supervised by the Ministry for Arms and Ammunition (*Reichsministerium für Bewaffnung und Munition*), which eventually centralized nearly all German wartime resources and war material production.[37]

When newly developed weapons, ordnance, or equipment were ready for submission to and testing by, for instance, the Army, the *HWaA* again stepped in via its subsidiary agency, the Office of Weapons Proof and Development (*Waffenamt Prüfwesen* or *WaPrüf*) to test, evaluate, and approve the design for production and issue.[38]

Chapter 10: Manufacturers' Logos, Markings, and Codes

Once put into general production, the *HWaA* tightly monitored production quality at the factories via the Inspections Office (*Abnahme* or *Heeres-Abnahmewesen*). The Inspections Office was:

"...responsible for the testing and acceptance of all weapons [and equipment]... before delivery to the *Wehrmacht*. Inspections were carried out according to detailed guidelines called *Tecnhische Lieferbedingungen* (TL) prepared by the various...(*WaPrüf*) departments."[39]

At the dawn of WWII, Germany's ordnance industry was geographically divided into eleven districts. Each district had a central ordnance depot (*Zeugamt*), and a team or teams of inspectors assigned by the *Abnahme*.[40] Each team had a supervisory inspector who was assigned a particular number, and issued a set of inspection dies (*Stempel*). These dies incorporated one or another variation of the national eagle, and the inspector's designated number. Supervisory inspectors may have been responsible for a single factory, or the output of an entire district. The supervisory inspectors were occasionally moved between districts, and simply took their number (and thus, their set of inspection dies) with them throughout the war.

Once a component or whole assembly met military requirements, the inspector would mark the item as acceptable. These *Abnahme* acceptance marks (*Wehrmachtabnahmestempel*) stamped into military items have become colloquially known in collector circles as "Waffenamts," or abbreviated "WaA." The fundamental difference between the stylized MPD marking, and that of the WaA is that the MPD marking is an identification of maker, material, and an implied *promise* from the manufacturer to the consumer that the item was made to DIN standards. The WaA stamp, on the other hand, is a positive *confirmation* that a particular item was made to military standards and deemed acceptable.

Items made from steel or leather were struck by the WaA inspector with a hardened die. Polymer plastics were marked in the same manner in many cases, but likely due to the possibility of shattering the item, markings have also been noted to have been heated or branded into the material, and some use of rubber stamping with paint has been noted.[41] Also, since some items were available to both civilians and the military, various combinations of codes and markings have been observed on like items.

Fig. 83: Examples of *Abnahme* acceptance stamps pertinent to the *Kunststoffe* industry, from left to right: common WaA 359 (Walther), WaA 285, WaA78, demonstrating variations in the method of marking, and the BAL marking as used by the *Luftwaffe*.

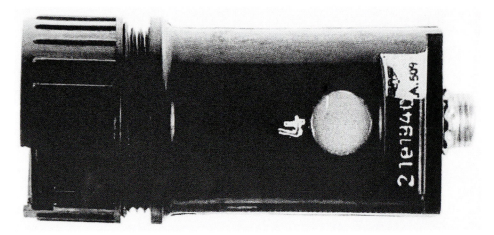

Fig. 84: A *Zt.Z.f. SpBü. 37* explosives timer with an ink stamped WaA 509 on white paint. *From US Army TM-E-30-451.*

Fig. 85: A pair of early P38 grips, bearing "480" and the early "Cewe" logo, both of which represented Walther, Zella-Mehlis, and WaA359 inspector markings. In addition, the grips are both molded with the code V7 for Walther in the upper MPD field. Interestingly, the grip on the left was made from T1, while the right grip was made from Z3 material. Courtesy *Leon DeSpain.*

Chapter 10: Manufacturers' Logos, Markings, and Codes

Fig. 86: A K98 bayonet grip plate made by Gebr. Spindler KG, Köppelsdorf, and stamped with the WaA155 by the area *Abnahme* inspector. *Courtesy Kenneth Huddle.*

Additional Designations and Markings

In the pre-WWII and early war periods, the German *Luftwaffe* appears to have had a good deal more control over the production of certain items it procured, and its own set of inspection and acceptance practices. This is particularly true of items that were components of or used onboard aircraft, such as precision flight and navigation instruments, housings, electrical components, and signaling equipment. Frequently, these items bear none of the codes discussed previously, but rather, are usually marked with at least an "FL" (*Fliegergerät Anforderungszeichen*), or "Aviator Equipment Requisition" number, and a "BAL" stamp (*Bauaufsicht Luft*), a *Luftwaffe* agency that served much the same function as the previously discussed *Heeres-Abnahmewesen*.

Other markings and information typically seen on *Luftwaffe* items include the abbreviations "*Anf*" (*Anforderungszeichen*, or standardized Ordering Number), "*Gerat Nr.*" (Device or Equipment Number), "*Werk Nr.*" (Serial Number), and/or the words "*Lieferer*" (Supplier) and "*Bauart*" (Type). These markings will generally be incorporated into the mold, but some stamping and ink/paint stamping has been noted, as well.

Fig. 87: A *Luftwaffe* phenolic *Patronenausstoßer 15* made by Gebrüder Spindler of Köppeldorf, designated Fl. 46480.

More comprehensive markings describing pressing material, percentage resin content, and color coding have also been described. The 1943 *Werkstoff Ratgeber* advised:

"The manufacturers of the pressing materials have agreed upon an identification of their products through trade numbers. These consist of the normed type designation and of a four-place number.

The first digit names the type of pressing material (*1* phenolic resin, *2* phenolic resin, ammonia-free, *3* phenolic resin, taste-free, *4* cresylic resin, *5* cresylic resin, ammonia-free). The second digit explains the resin content (*3* 35%, *4* 40%, *5* 45%, *6* 50%, *7* 55%, *8* 60%, *0* 100%). The two last digits point to the color. The following is signified: *00 to 09* white, ivory, yellow, natural, *10 to 19* brown, *20 to 29* pink, red to mahogany, *30 to 35* green, *36 to 39* blue, *40 to 49* gray to black, *50 to 79* marbleized, *80 to 99* spotted...."[42]

The author has not noticed widespread compliance with this guidance, and only infrequently have these additional resin content and/or hue designators been noted on wartime items.

They can be confused with centralized requisition numbers, manufacturer's mold/product identification numbers, or mold position numbers, which are much more commonly encountered. Product numbers appear to have been used internally by companies involved in producing hundreds or thousands of different types of items to identify which mold to pull from storage for a particular run, or to determine exactly which mold may have problems or defects by examining the end product. So, too, these mold/product numbers may have been used by some firms for cataloging and sales orders. Mold position numbers are seen on smaller items made in a multiplex mold, again probably used internally by the company to determine the exact position of defects in the mold, or to ensure that a particular right and left pair of grip plates, for instance, remain together, and thus ensuring continuity in construction and hue.

Fig. 88: MPD marking, whereas M4 represents Kronacher Porzellanfabrik Stockhardt und Schmidt-Eckert, Kronach, and the unusual "31/131" in the material classification field, which presumably represents the type of pressing material, resin content, and resin color. *Courtesy Kenneth Huddle.*

Chapter 10: Manufacturers' Logos, Markings, and Codes

Fig 89: A late WWII steel canteen painted with red oxide enamel, and fitted with a green phenolic canteen cup.

Fig. 90: Green canteen cup marked with the military letter code for Julius Posselt, Gablonz a. Neiße (Reichsgau Sudetenland) "gfc," but the variant "31" has been used instead of the traditional material classifications.

Fig. 91: A phenolic carbide lantern spare parts box with sturdy metal clasp, made by Presswerk-AG, Essen.

Fig. 92: Inside of the same carbide lantern spare parts box. The company molded in their logo, as well as their MPD code (omitting the material classification designator). The number "1658" represents "phenolic resin, 50% resin content, marblized." Sequential numbers or letters tend to indicate molds used for parts that would normally go together, such as tops and bottoms, right and left grip plates, lids and main container etc. In this case, "/A" designates the lid to the hinged container, which is identified as "/B."

Chapter 10: Manufacturers' Logos, Markings, and Codes

Fig. 93: Inside of a brown *Kunststoff* binocular case marked "ehe" and "41," a variant material classification designator. The number 9951 is presumed to be a centralized requisition number.

11

Selected Notes on Common Wartime Products

By far, the most numerous WWII German *Kunststoff* items that will be encountered by the collector or enthusiast are personal items, such as small arms oilers, "butter dishes," anti-chemical agent containers, and various examples of ordnance and fuse shipment caps and containers. Some notes concerning these items follow.

Most soldiers were issued with a *Reinigungsgerät 34*, or "Cleaning Equipment 34," commonly called the "tobacco can cleaning kit" by collectors, to perform small

Fig. 94: Early machined oiler (left), the widely issued phenolic type (center, left), cast pot-metal (center, right), and an example of the sheet metal "stamped" oiler (far right). These later oilers were developed to further simplify RG34 production.

Chapter 11: Selected Notes on Common Wartime Products

arms maintenance. Initially, the kits were issued with a machined steel oiler as one of its components. These were labor-intensive to produce, and thus a phenolic pattern was developed and introduced circa 1941-42. Produced in the hundreds of thousands, if not millions, many collectors are unaware that components of the oilers are frequently found marked with tiny MPD codes or military letter codes.

Fig. 95: The RG34 phenolic oiler disassembled.

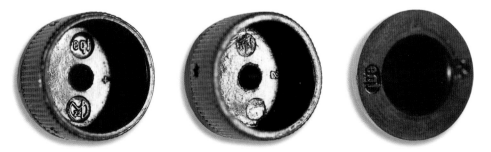

Fig. 96: Three examples of markings that can be found within the RG34 oiler.

Before and during the early years of the war, each soldier was reportedly issued four small phenolic skin decontamination tablet (*Hautentgiftungsmittel*) containers. Each container was packed with ten tablets of "Losantin," which when prepared with water was used to decontaminate the skin. The brown *Kunststoff* containers were sealed with adhesive tape, the color of which indicated the year of manufacture (red tape used up to 1940, black tape represented 1941, light green tape 1942, and yellow tape meant 1943).[43] The original program and contracts called for sixty million (60,000,000) of these containers to be made, but it is doubtful that German industry produced anywhere near that amount.[44]

Fig. 97: An example of a brown container made to hold *Hautentgiftungsmittel* tablets, made by dozens of manufacturers and widely available on the current militaria market.

Fig. 98: *Hautentgiftungsmittel* made in 1943, and marked "jle" for Établissements Alfred Barth, Brüssel (Bruxelles), Belgium, a company working under German occupation.

Chapter 11: Selected Notes on Common Wartime Products 93

Since mixing the tablets with water in the midst of a chemical attack was, of course, less than ideal, an easier to use pre-mixed ointment was developed for skin decontamination. The small bottles made to hold the *Hautentgiftungssalbe*, or skin decontamination salve, were made from four separate molded parts, and can be found in orange, brown, and brown-black hues. They were made by a myriad of manufacturers, from both phenolic resins as well as polystyrene.

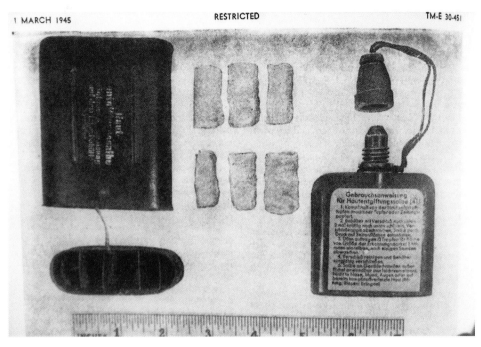

Fig. 99: The complete German wartime chemical agent decontamination kit issued to each soldier, including the oft encountered "*Hautentgiftungssalbe*" salve bottle. *From US Army TM-E-30-451.*

Fig. 100: The base of a "*Hautentgiftungssalbe*" salve bottle made in 1944 coded "ggh," representing Robert Scneider KG, Offenbach am Main.

Kunststoffe

Fig. 101: An assortment of M22 and M28 *Flak* and *Pak* projectile shipping plugs.

Fig. 102: Note the variations and manner of marking in just this small sampling of six ordnance shipping plugs.

Chapter 11: Selected Notes on Common Wartime Products

Projectile shipping caps, as well as ammunition, fuse, and ordnance containers are also plentiful currently, and found in many varieties. Most will have some sort of identifying term or abbreviation molded in, which can be confusing for those that do not read German, or understand the methodology of German abbreviations.

"*M22 x 1,5*," "*Ersatz für den Leuchtspureinsatz.*" These little twist on caps were used to protect the threaded tracer element receptacle of 3.7 cm and 5 cm *Flak* or *Pak* projectiles during transit or shipping. These plugs were removed and discarded when the fuse and projectile were readied for firing, and can sometimes be found with dates and mold position numbers on the reverse. The numbers "M22 x 1,5" on these and the following products represent the metric thread dimensions.

"*M22 x 1,5 z*," "*Ersatz für den Pz. Gr. Zünder*," or "Replacement for the Armored Piercing Projectile," which was used to protect the vulnerable threads on the base of these 5 cm rounds during shipment; also meant to be discarded.

"*M28 x 1,5*," "*Ersatz für den Kopfzünder*," or "Replacement for the Nose Fuse." Projectiles were shipped separately from fuses for safety. To protect the threads in these projectiles, phenolic plugs were inserted during shipment. This type was used to protect the nose fuse threads of 37mm *Flak* and *Pak* projectiles, as well as those fired in the 5 cm Pak K.W.K., during shipping, and can sometimes be found with dates and mold position numbers on the reverse. A plethora of these small shipment plugs are to be found on the current market.

"*2L*" or "*2L verl.*" These are containers for aerial bomb fuses as used by the *Luftwaffe*. A number of manufacturers have been noted, and all observed examples have been made of "S" material (phenolic resin with fine sawdust filler). The "*verl.*" abbreviation indicated a lengthened container.

Fig. 103: Deep brown aerial bomb fuse container marked "2L verl." and "eje" inside (Blumberg & Co. Bürobedarfsartikelfabrik).

"*gr. Zdlg. C/98 o.V.*," "*grosse Zündladung C/98 ohne Verzögerung*," or "large gaine/booster C/98 (Construction 1898) without delay." These slim two-piece containers held five each, and can be found in light tan-brown or deep brown hues. A number of manufacturers have been noted, and all observed examples have been made of "S" material (phenolic resin with fine sawdust filler).

"*kl. Zdlg. C/98 o.V.*," "*kleine Zündladung C/98 ohne Verzögerung*," or "small gaine/booster C/98 (Construction 1898) without delay." These slim two-piece containers held five each, and can be found in light tan-brown or deep brown hues. A number of manufacturers have been noted, and again, all observed examples have been made of "S" material.

Fig. 104: A "*gr. Zdlg. C/98 o.V.*" container made by Josef Mellert, Bretten.

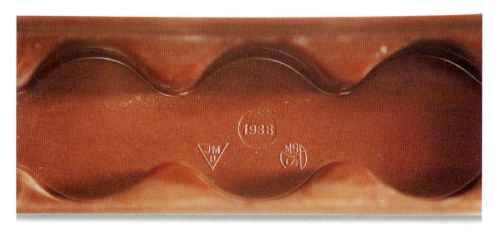

Fig. 105: Inside close up of the same "*gr. Zdlg. C/98 o.V.*" container made by Josef Mellert, Bretten. Note the JMB inside a triangle logo and the MPD marking "M9," indicating the firm, and "S" the material (phenolic resin with fine sawdust filler).

Chapter 11: Selected Notes on Common Wartime Products

"*Leuchtpatronenbehälter*," or "Flare Cartridge Holder." The *Wehrmacht* used various patterns of single and double barreled flare pistols for signaling, illumination, and marking targets. Fluted carriers were produced with twist off lids to protect six of these 26 mm flares. The container was to be turned in and reused.

Fig. 106: A brown phenolic *Leuchtpatronenbehälter* as seen from the side. Presumably, once the flares were used, the containers were turned in and sent back to Germany to be filled by munitions companies, and returned to the front again.

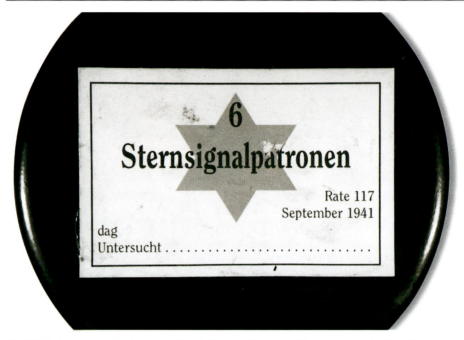

Fig. 107: Various paper labels can be found glued to the top of flare cartridge holders. Once back in Germany, the containers were probably cleaned, filled up, and new labels applied. This one indicates that the contents include six red star signal cartridges made in 1941 by the firm Orion Metallwarenfabrik GmbH of Kremmen, as indicated by the ordnance code "dag."

Fig. 108: A collection of the larger *Kartusch Vorlage* containers. *Courtesy Kenneth Huddle*

Chapter 11: Selected Notes on Common Wartime Products

"*Kart. Vorl. Geb. G. 36*" is an abbreviation of *Kartusch Vorlage Gebirgsgeschütz G.36*, or roughly "flash reduction charge for the Mountain Cannon G.36." These containers held charges that helped decrease the muzzle flash of the small portable 75 mm G.36 cannon that could be broken down into 11 mule pack loads.

"*Kart. Vorl. H.13*" is an abbreviation for "*Kartusch Vorlage Haubitze 13*," or roughly "flash reduction charge Cartridge for the Howitzer 13." These 17 cm x 16 cm *Kunststoffe* containers held several charges each for use with the s.F.H. 13 heavy field gun.

"*Kart. Vorl. l. F.H. 16/18*" is an abbreviation for *Kartusch Vorlage leichte Feldhaubitze 16/18*, or roughly "flash reduction charge for the light Field Howitzer 16/18." The two main field howitzers in most WWII German artillery battalions were the Light Field Howitzer 16 (developed during WWI) and the Light Field Howitzer 18, introduced and fielded in 1935, which gradually replaced (though not completely) the earlier model.

"*Kart. Vorl. s.Gr.W.34*" is an abbreviation for *Kartusch Vorlage schwere Granatwerfer 34*, or roughly "charge Cartridge for the Heavy Mortar 34." Used with one of the *s.Gr.W.34* 80 mm mortar fielded before and during the war.

"*Sprengkapseln (Duplex)*," or "Gaine/booster (Duplex)," seen on small two piece phenolic boxes meant to hold gaines/boosters for 20 mm projectiles. The designation "*Duplexkapseln*" has alternately been observed molded into these boxes.

"*Zt. Z.f. Sp Bü. 37.*" is an abbreviation for "*Zeit-Zünder für Sprengbüchse 37*," a timed fuse for use with demolitions explosives.

"*Zünderbüchse 1*," or "Fuse container 1." These common tapered conical containers held artillery fuses. Widely available after the war, many were made by savvy civilians into salt and pepper shakers by drilling a number of holes through the screw-off lid. "*Zünderbüchse 2*" marked containers have also been observed, but vary little from the aforementioned.[45]

Fig. 109: Left and bottom views of the commonly available fuse *Zünderbüchse 1* fuse container.

Chapter 11: Selected Notes on Common Wartime Products

Fig. 110: A collection of the many *Zünderbüchse 1* and *Zünderbüchse 2* fuse containers available, demonstrating the different hues that can be encountered. *Courtesy Kenneth Huddle.*

Fig. 111: An interesting finned shipping container made in 1941 by "ehj" (Gebr. Betr. Spindler K.G. of Köppelsdorf).

Fig. 112: Same finned shipping container shown open. The exact piece of ordnance this container protected is not known.

12
Selected Topics and Interesting Military Products Made during WWII

Some amazingly complex and enormous presses and molds were developed by the German plastics industry during the war. Larger items, such as thermoplastic suitcases, large toolboxes, and furniture are of course taken for granted these days, but in the 1940s, anything made from thermosets larger than a shoebox was quite an accomplishment.

Fig. 113: A Canadian investigator examining a large phenolic battery box molded by Bisterfeld & Stolting. From *"German Plastics Practice,"* Office of the Quartermaster General, 1945.

Experimental snow skis and long, large, complicated housings, such as complex battery box moldings, were produced. Equally impressive was a model of heavy bearings made by Bisterfeld & Stolting during the war. The firm developed a special thermoset material called "Type 11," and claimed that the lifespan of the bearings was greater than those made from bronze, provided that loads were not too heavy, and the speeds too high. These bearings were presumably intended for armored vehicle or maritime applications.

Fig. 114: A huge, multi-piece bearing molded by Bisterfeld & Stolting. From "German Plastics Practice," Office of the Quartermaster General, 1945.

Fig. 115: A rare set of "*Dienstglas 6x30*" binoculars made almost entirely of phenolic resin by Emil Busch, Rathenow. These models ironically came from the factory in pressed-paper *ersatz* leather cases, instead of the expected hard *Kunststoff* type so frequently encountered.

Chapter 12: Selected Topics and Interesting Military Products

Fig. 116: Same "cxn" marked pair of binoculars, serial number 417619. The few metal fittings were painted *Wehrmacht* beige.

Fig. 117: Same "*Dienstglas 6x30*" as seen from side and from top. The military letter code "cxn" represented Emil Busch A.G., Optische Industrie, Rathenow.

Manufacturers of optical equipment, as has been discussed, used quite an amount of *Kunststoffe* during WWII for eyepieces, cases, and components. One firm (Emil Busch of Rathenow), however, took the concept one step further, creating a set of binoculars made almost entirely of phenolic resin. The few metal components were painted *Wehrmacht* beige and, other than the glass prisms and lenses, the rest of the assembly was made from an attractive red-brown pressing material.

Germany had an acute shortage of seasoned hardwoods for small arms furniture, especially walnut, due to its rapid rearmament. To cope, arms manufacturers explored a number of alternatives, such as oak and poorer and softer phenolic-bonded, beechwood (which was and is plentiful in Central Europe) laminates for items such as rifle stocks. But even if the country had a plentiful supply of walnut, items made from solid wood are subject to expansion, contraction, and damage from the elements, or from day to day field use. Those made from laminated wood and phenolic adhesives might suffer these effects somewhat less severely, but they were heavier, and took more effort and expense to produce.

One possible solution to the problem was the idea of synthetic or plastic stocks. To this end, a small number of K98k rifles made by Mauser Werke, Borsigwalde, were fitted with composition furniture for troop trials, sometimes mistakenly called "fiberglass" or "Bakelite stocks" by collectors. These terms are, however, incorrect.

All contemporary reports and post-war literature describe these rifle stocks as having been made from one long layer of fabric, formed to the same shape as a traditional K98k stock, and all held together with a synthetic polymer resin. One source states that the polymer used to bind this altogether was urea-formaldehyde, as opposed to phenol-formaldehyde, but this could not be confirmed.[46] The end result was a strong, hollow formed stock, finished with what is described as a red lacquer, and then reinforced with metal inserts at key stress and wear areas.[47]

Other than the peculiar stock, the rest of the rifle and components are of the standard K98k variety. Of the rifles observed, all were stamped with the same manufacturer's codes, either S/243 or simply 243, and dated either 1937 or 1940. Most of the stocks observed were stamped with the weapon's serial number, and on the right butt, an eagle over an "H" (*Heer*), and one or two early "droop eagle" WaA217 acceptance stamps.

The result of testing and trials must not have been as favorable as envisioned, or the pace of the war simply overtook the idea, as large numbers were apparently never fielded, and few of these synthetic stocked K98ks survived the war.

Chapter 12: Selected Topics and Interesting Military Products

Fig. 118: An example of a K98k housed in a polymer composition stock. *Leon DeSpain collection*

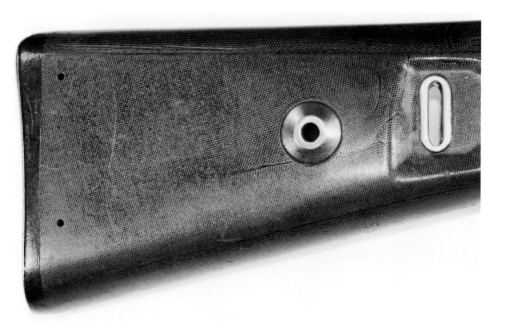

Fig. 119: Close up of the right butt of same composition polymer stock. This example is proofed with Eagle over "H," one large WaA217, and another smaller on the side, and the weapon's serial number can be seen stamped on the bottom. The hollow stock was closed at the rear with a phenolic or ureic cap, pinned in place, and a standard steel butt plate was screwed in behind. Note the aluminum reinforcement of the sling slot (a high wear and stress area). *Leon DeSpain collection*

Fig. 120: Another example of a K98k housed in a polymer composition stock, again marked WaA 217 and Eagle/"H," indicating issue to the *Heer*. *Robert Jensen collection*

Fig. 121: Collectors and WWII enthusiasts should not minimize nor forget that certain *Kunststoff* items available today had ominous connections. In this case a container for "*Zyklon-B*," the poison gas used in the concentration camps to murder untold numbers of men, women, and children. *Zyklon B* was originally developed as a pesticide by the infamous I.G. Farben (a mammoth petro-chemical-pigment conglomerate formed in 1925 by merging the Bayer, AGFA, and BASF corporations with two smaller companies).[48]

13

Allied Appraisal of the Wartime German Plastic Industry

Perhaps the makers of *Kunststoffe* should have taken the safeguarding and cloaking of the corporate identities and locations a little more seriously from the start. Some were of course too well known already, or too big to escape the eye of Allied intelligence officers. The result was that many of the larger firms had indeed been targeted and bombed by the Anglo-Americans by the end of the war.

In Berlin, Bakelite G.m.b.H itself was reported to have been completely destroyed from the air in 1943.[49] Dynamit A.G., Troisdorf, was also hit so severely on December 29, 1944, that it was forced to cease *Kunststoffe* production completely.[50] I.G. Farben, Ludwigshafen, and Kurt Albert G.m.b.H., Wiesbaden, were also severely damaged.[51]

Industries in entire regions were in some cases simply obliterated. The casein industry, for instance, was already crippled in 1941 from a lack of raw materials, but the Allies still targeted the area. One report listing the status of the smaller firms stated:

"[International Galalith-Gesellschaft:] ...very extensively damaged. With a few exceptions...the plant has been either destroyed completely or it has been damaged by blast, fire, water and subsequent exposure...

O. Oswald & Co., Hamburg	Missing
Karl Probst, Nuremberg	Closed down
Karl Schwab, [Nuremberg]	Destroyed
Sirolith G.m.b.H., Berlin	[Destroyed]
Hannoverische Kunsthornfabrik, Busch & Co.	[Destroyed]"[52]

Many of the smaller firms did escape the war without being bombed or damaged too severely. Once order had been restored, and when they were able to secure enough raw materials, molding powders, labor, and electricity, a fair number of companies resumed operations and shifted production to consumer goods. It also didn't take a few of the larger firms long to get some of their production lines back into operation.

Even before the close of hostilities, as the Allies advanced deeper and deeper into the Reich, SHAEF sent out teams of technical evaluators to investigate German military and industrial targets of interest, which culminated in the invaluable European Technical Intelligence Reports, and the subsequent FIAT, BIOS, and CIOS reports referenced throughout the text.

In general, the Allied appraisal of the German plastics industry varied. On one hand, the Allied experts were amazed at some of the processes and techniques developed during the war, especially in regard to the development of newer thermoplastics such, as PVC, synthetic rubber, corrosion resistant pigments, rot resistant fiber from chlorinated polyvinyl chloride, and others.[53]

"Their chemical research has been impressive..."[54]

On the other hand, investigators did not find the manufacturers to be very efficient or organized in comparison to firms in, say, the United Kingdom or United States. The opinion of the various investigators was that German industry had not really made any revolutionary advancements during the war with thermosets. Furthermore:

"...their fabrication and molding operations, while they involve ingenious devices, are essentially small scale as compared to American standards."[55]

And, though the *Kunststoffe* industry was able to turn out an estimated 30,000 tons of phenolic molding powders per year:

"...the quality of German phenol and cresol moulding materials fell off about the middle of 1942. About the same time cotton fillers were replaced with paper. The resin content of these powders was reduced, in some cases to about 30%."[56]

14

Post-War Miscellany

Most of the thermoset goods produced just after the war in the Western zones of occupied Germany were for industrial use, or of the consumer goods variety. Of those observed by the author, most bear only manufacturers' logos, mold numbers and the like, and it appears that the use of the MPD marking either fell from favor, or disappeared immediately after the war. It is presumed that some firms would have continued using it at least as an interim measure, but the author has not observed the marking on any item confirmed to have been produced during the 1945-1955 occupation period.

The few *Kunststoffe* concerns in the Soviet zone of occupation that survived the war were eventually absorbed by the fledgling state, the German Democratic Republic, or *Deutsche Demokratische Republik* (DDR). All of these companies were reorganized and renamed as state-owned properties as indicated by the VEB, or *Volkseigener Betrieb* ("People's Owned Company") designator, used in place of the familiar "A.G.," "K.G.," or "G.m.b.H." corporate designations. Of note, a modification of the wartime MPD marking containing only a stylized "P" appears to have been used in the DDR in the product moldings.

The use of thermosets as the preferred, or at least most common molding materials continued in the DDR, and indeed the rest of the Warsaw Pact for many decades. In the west, however, the manufacture of goods from "low tech" thermosets quickly fell from favor, largely replaced by the use of more advanced thermoplastics, such as urethanes, butyrates, ethyls, vinyls, acrylics, polystyrenes, nylons, and polyethylenes.

Fig. 123: The MPD marking as modified and molded into products thought to have been made in the DDR. *Courtesy Kenneth Huddle*

Fig. 122: A post-WWII fuse container thought to have been made in the DDR (left), and the standard wartime model (right). Note how the post-war version "fuze tip locating cup" is molded in place, rather than being a separate piece inserted inside. *Courtesy Kenneth Huddle*

Appendix I:
Material Classification Reference

In numeric and alphabetical order, as cataloged and summarized by the author:

Type	Description	Type of Filler Material and Remarks
0	Phenolic resin with wood filler	Sawdust; composition for slow pressing
1	Phenolic resin	
2	Phenolic resin, ammonia-free	
3	Phenolic resin, tasteless	
4	Cresylic resin	
5	Cresylic resin, ammonia free	
6	Natural resin or Bitumen	Inorganic filler
7	Natural resin or Bitumen	Inorganic filler
8	Bitumen	Inorganic filler
11	Phenolic resin with inorganic filler	Rock dust (presumably talc)
12	Phenolic resin with inorganic filler	Asbestos fibers
A	Cellulose Acetate	"Celluloid"
K	Urea resin with organic filler	e.g., cellulose [as filler material]
M	Phenolic resin with inorganic filler	Asbestos cord
S	Phenolic resin with wood filler	Sawdust
T1	Phenolic resin with textile fiber filler	Short textile fibers
T2	Phenolic resin with textile fiber filler	Shredded textile fabric
T3	Phenolic resin with textile fiber filler	Textile fabric strips
X	Silica cement compound	Asbestos or alternate organic filler
Y	Lead and ash filler	
Z1	Phenolic resin with cellulose filler	Short-fibered cellulose (flakes)
Z2	Phenolic resin with cellulose filler	Cellulose shreds (e.g., paper shreds)
Z3	Phenolic resin with cellulose filler	Layered cellulose (e.g., paper strips)

Appendix II: Manufacturers by Observed Trademarks/Logos

Many of the *Kunststoffe* manufacturers incorporated their unique and proprietary logos and trademarks into the molds and products they produced. Those that can be firmly established or connected to one of the traditional plastics manufacturers follow, listed in alphabetical order. As there were hundreds of companies involved in *Kunststoffe* production (as prime and sub-contractors), and tens of thousands of different articles produced, this list should in no way be considered complete. Numbers in brackets are the known MPD numerical or alphanumerical identifier, or the known military letter code for the firm.

Acla, Rheinische Maschinen-, Leder- u. Riemenfabrik, Aktiengesellschaft, Köln-Mülheim [OZ]

Emil Adolff, Abt. Kunstharz-Presswerk, Reutlingen i. Württ. [H2, "cuy"]

Allgemeine Elektricitäts-Gesellschaft, Fabriken Hennigsdorf [38, "enl"]

Gebr. Berker, Schalksmühle i. Westf. [80, "ggl"]

Gebr. Berker, Schalksmühle i. Westf. [80, "ggl"]

Bakelite Gesellschaft G.m.b.H, Erkner bei Berlin [doesn't appear to have been assigned a number or letter code]

BEBRIT-Preßstoffwerke G.m.b.H., Bebra u. C. & F. Schlothauer G.m.b.H., Ruhla (Thür.); Bebra (H.-N.) [22]

Robert Bosch G.m.b.H., Metallwerk, Stuttgart-Feuerbach [55, "mpm"]

Robert Bosch G.m.b.H., Metallwerk, Stuttgart-Feuerbach [55, "mpm"]

Appendix II: Manufacturers by Observed Trademarks/Logos   *115*

 Max Braun, Frankfurt a. Main [3A]

Gebr. Broghammer, Schramberg (Schwarzwald) [F8]

 Busch-Jaeger Lüdenscheider Metallwerke Aktiengesellschaft, Lüdenscheid i. Westf. [23, "xa"]

 Dynamit-Actien-Gesellschaft vormals Alfred Nobel & Co., Abteilung Celluloid-und-Kunstoff-Fabrik, Werk Troisdorf (Bez. Köln); verkauf durch Venditor, Kunststoff-Verkaufsges. m.b.H, Troisdorf (Bez. Köln) [43, "boa"]

 Dynamit-Actien-Gesellschaft vormals Alfred Nobel & Co., Abteilung Celluloid-und-Kunstoff-Fabrik, Werk Troisdorf (Bez. Köln); verkauf durch Venditor, Kunstoff-Verkaufsges. m.b.H, Troisdorf (Bez. Köln) [43, "boa"]

 Christian Geyer, Nürnberg [83, "ajs"]

 thought to represent Grelit, Kunstharz-Presswerk Grohmann, Pietschmann & Co., Nixdorf (Reichsgau Sudetenland) [1Y, "cws"]

 Heliowattwerke, Elektrizitäts-Aktiengesellschaft, Berlin-Charlottenburg [35, "hpx"]

 Isola-Werke A.G., Birkesdorf -Düren (Rhld.) [40, "bdl"]

Paul Jatow, Dodendorf (bez. Magdeburg) [E4]

 Köditz & Co., Metallwarenfabrik u. Presserei, Langeweisen i. Thür. [Z9]

Hugo Krieger und [Hermann] Faudt, Berlin [78, "etl"]

116 Kunststoffe

Kronacher Porzellanfabrik Stockhardt und Schmidt-Eckert, Kronach i. Bayern [M4, "erz"]

"List" brand, seen on *Luftwaffe* electrical components, exact maker unconfirmed, but thought to be Heinrich List Elektro-Meßgerätebau, Berlin-Steglitz [4Z, "hbu", "ocw"]

C. Lorenz Aktiengesellschaft, Berlin-Tempelhof [Y9, "dmr"]

Meirowsky und Co. Aktiengesellschaft, Porz a. Rhein [L4, "ggf"]

Jos. Mellert, Bretten i. Baden [M9, "bzh"]

Radio H. Mende & Co., Abt. Presswerk "Mendelith", Dresden [56, "bl"]

Radio H. Mende & Co., Abt. Presswerk "Mendelith", Dresden [56, "bl"]

Gebrüder Merten, Gummersbach [24, "gdr"]

Mix & Genest Aktiengesellschaft, Berlin-Schöneberg [68, "bbc"]

New York-Hamburger Gummi-Waaren Compagnie, Abt. Kunststoffe, Hamburg [21, "bfn"]

Phenoplast, Bischoff & Co., Kom.-Ges., Eberswalde [V1, "egx"]

Philips; Deutsche-Philips Gesellschaft m.b.H., Berlin [58]

Appendix II: Manufacturers by Observed Trademarks/Logos   117

Presswerk A.G., Essen [45]

Dr. F. Raschig G.m.b.H., Chemische Fabrik, Ludwigshafen a. Rhein [Ra]

Richard Rinker G.m.b.H., Menden (Krs. Iserlohn) [X0, "brb"]

[RWS] Rheinisch-Westfälische Sprengstoff Actiengesellschaft, Werk Stadeln, bei Fürth [dnf], (acquired later by Alfred Nobel & Co, then in turn even later by Dynamit A.G., [43])

[RWS] Rheinisch-Westfälische Sprengstoff Actiengesellschaft, Werk Stadeln, bei Fürth [dnf], (acquired later by Alfred Nobel & Co, then in turn even later by Dynamit A.G., [43])

H. Römmler Aktiengesellschaft, Spremberg Nd.-Lausitz
[HRS, 32, "ehe"]

Pressstoffwerk Schöppenstedt, Paul Schnake, Schöppenstedt
[81, "epy"]

Siemens-Schuckertwerke, Aktiengesellschaft, Abteilung Isolierstoffe (SK 4), Berlin-Siemensstadt (Gartenfeld) [34]

Süddeutsche Isolatoren-Werke G.m.b.H, Freiburg im Breisgau [25, "jnf"]

Thiel & Schuchardt, Metallwarenfabrik Gesellschaft mit beschränkter Haftung, Ruhla i. Thür. [L3, "edv"]

118 *Kunststoffe*

Hein[rich]. Ulbricht's Witwe., Ges.m.b.H., Kaufing bei Schwanenstadt, Oberdonau [3X, "ans"]

Vossloh-Werke G.m.b.H., Werdohl i. Westf. [E2, "flf"]

Carl Walther Waffenfabrik, Zella-Mehlis, Thüringen [V7, "ac"]

Carl Walther Waffenfabrik, Zella-Mehlis, Thüringen [V7, "ac"]

Carl Walther Waffenfabrik, Zella-Mehlis, Thüringen [V7, "ac"]

Elektrotechnische Fabrik Weber & Co., Komm.-Ges., Kranichfeld i. Thür. [98, "epz"]

Likely: Dr. Wellborn & Wernicke, Berlin [0A, "gbk"]

Casp[ar]. Arn[old]. Winkhaus, Carthausen i. Westf. [H1, "adh"]

The following short list includes some observed logos incorporated into molded *Kunststoffe* products prior to or during the war, but the actual pressing material and/or processing thought to be contracted to one of the traditional plastics firms. The author has attempted to limit these to only those products and firms that are confirmed to be of pre-WWII and wartime vintage. Again, as thousands of products were manufactured during this era, this list should not be considered complete.

Askania-Werke, Berlin-Friedenau [kjj]

AGFA; I.G. Farben-Industrie AG, Werk "Agfa", Berlin [mbv], München [bzz]

Appendix II: Manufacturers by Observed Trademarks/Logos 119

 Blaupunkt-Werke G.m.b.H., Berlin-Wilmersdorf [fvw]

 Brandt, Roland, Fabrik für Radiotelefonie, Berlin [bq]

 (unconfirmed trademark)

 Emil Busch A.G., Rathenow [cxn]

 Not a manufacturer's trademark but a representation of the *Verband Deutscher Elektrotechniker* (German Federation of Electrical Engineers)

 DUX G.m.b.H., Präzisions-Bleistiftspitzer, Hemhofen

 Erfurter Maschinenfabrik Berthold Geipel GmbH (ERMA) [ayf]

 (unconfirmed trademark)

 (unconfirmed trademark)

 Geha-Werke G.m.b.H., Hannover

 Graetz (radio manufacturer)

Grundig A.G., Fürth

Hartmann & Braun AG, Elektrische & wärmetechnische Meßgeräte, Werk Braunschweig [jnm]; Frankfurt [fpb]

[Hella] Westfälische Metall Industrie KG Hueck & Co., Lippstadt [dom]

(unconfirmed trademark)

(unconfirmed trademark)

(unconfirmed trademark)

(unconfirmed trademark)

KACO; Gustav Bach Kupfer-Asbest-Co., Heilbronn [eqq]

KAW (unconfirmed trademark)

(unconfirmed, maker of compasses with phenolic housings marked "T6", Elektro-, Glimmer und Preßwerke Scherb & Schwer K.-G. vorm. Jareslaw, Berlin-Weißensee)

KUM & Co., K.G.., Kunststoff- und Metallwarenfabrik, Erlangen (thought to have been a wartime manufacturer as well as post-WWII)

Ernst Leitz, G.m.b.H., Optische Werke, Wetzlar [beh]

Appendix II: Manufacturers by Observed Trademarks/Logos

Gesellschaft f. elektr. Unternehmungen AG: Fabrik Loewe-Gesfürel AG, Berlin [fqv]

Luftgerätewerk Hackenfelde G.m.b.H., Berlin-Spandau [nhu]

(unconfirmed trademark)

Mauser-Werke AG, Oberndorf a.N. [S/42, 42, byf, svw]

NORA; Siemens-Electrogeräte G.m.b.H., Berlin [Berlin Charlottenburg]

NOTEK, Nova-Technik G.m.b.H, München [esq]

OSRAM G.m.b.H., Glühlampenfabrik, Berlin [mqe]

Not a manufacturer's trademark but a representation of the Österreichischer Verband für Elektrotechniker (Austrian Association of Electrical Engineers)

"Pelikan" Günther Wagner, Hannover [eor]

Schwarzwälder Apparate Bauanstalt (SABA), Villingen [dnz]

Siemens & Halske A.G., Abt. Luftfahrtgeräte, Berlin-Siemensstadt [eas, mqc]

Optische Werke C. A. Steinheil Söhne G.m.b.H., München [bmt, esu]

Süddeutsche Telefon,-Apparate,- Kabel und Drahtwerke A.G., Nürnberg [bug]

Süddeutsche Telefon,-Apparate,- Kabel und Drahtwerke A.G., Nürnberg [bug]

Telefunken Gesellschaft für drahtlose Telegraphie m.b.H., Werke-Erfurt & Berlin-Zehlendorf [bou]

Telefunken Gesellschaft für drahtlose Telegraphie m.b.H., Werke-Erfurt & Berlin-Zehlendorf [bou]

Ulbricht´s Witwe GmbH, Metallwarenfabrik, Kaufing, bei Schwanenstadt [ans]

Voigtländer & Sohn AG, Optisches Gerät, Werk Brauschweig [ddx]

Volkswagenwerk G.m.b.H., Stadt des KdF-Wagens bei Fallersleben [eky]

Gebr. Winter, Jungingen

Westfälische Metallwarenfabrik GmbH, Iserlohn (lkd), but also thought to have been used after the war in the DDR to represent "VEB Werkzeugmaschinen und Werkzeuge"

Zeiler; Batterien- u. Elementen-Fabrik, System Zeiler AG, Berlin [fef], a maker of flashlights, subcontracted/used phenolic components from H. Römmler Aktiengesellschaft, Spremberg Nd.-Lausitz [HRS, 32]

Carl Zeiss, Optische Geräte / Militärabteilung, Jena [blc, rln]

Appendix III:
Abbreviations and Company Initials

The first part of the 1940 article *Bekanntmachung über Kunstharz-Preßmassen für typisierte und überwachte Preßstoffe* (Announcement Concerning Synthetic Resin Molding Compounds for Standardized and Inspected Moldings) as it appeared in *Kunststoffe* Bd.30, Heft 3, Table I, pgs. 80-81, is transcribed, translated (by Paul Seibold), and provided for the reader below.

After the introduction, the original document listed established manufacturer's initials and abbreviations. It also had columns listing the predominant trade names and material types for each firm, which is not provided here. Abbreviations and their manufacturers as extracted and excerpted from *Bekanntmachung über Kunstharz-Preßmassen für typisierte und überachte Preßstoffe* follow the introduction. Additional abbreviations listed in brackets are from the author's observations and cross-referencing during the course of study and research. All are listed in alphabetic order for reference, as cataloged and summarized by the author:

"State Material Testing Office in Berlin-Dahlem
Unter den Eichen 87
Telephone: 76 27 51

Announcement Concerning Synthetic Resin Molding Compounds for Standardized and Inspected Moldings [57] [58]

According to the agreement between the State Material Testing Office in Berlin-Dahlem and the Technical Alliance of Manufacturers of Standardized Molding Compounds and Moldings (Registered Association), the synthetic resin molding compounds listed in Tables A, B and C are continually inspected by the Office. Only these synthetic resin molding compounds may be introduced into commerce with the designation "standardized and inspected by the State Material Testing Office in Berlin-Dahlem." The packaging of these molding compounds bears the inspection marking [MPD] (see DIN 7702), in which the firm's logo and the type symbol are given.

Molding facilities which do not themselves produce synthetic resin molding compounds may use only the synthetic resin molding compounds listed in Tables A, B and C for manufacturing synthetic resin molded parts (synthetic resin moldings) with the inspection marking.

Those synthetic resin molding compounds which as moldings satisfy the special electrical requirements listed in the standardization and bear a star in addition to the type symbol are specially listed in Table C. [59]

Staatliches Materialprüfungsamt in Berlin-Dahlem
Unter den Eichen 87
Fernsprecher: 76 27 51
Bekanntmachung über Kunstharz-Preßmassen[1]
für typisierte und überwachte Preßstoffe[1]) [2])

Gemäß Vertrag zwischen dem Staatlichen Materialprüfungsamt Berlin-Dahlem und der Technischen Vereinigung der Hersteller typisierter Preßmassen und Preßstoffe e. V. werden vom Amt die in den Tafeln A, B und C aufgeführten Kunstharz-Preßmassen ständig überwacht. Nur diese Kunstharz-Preßmassen dürfen mit der Bezeichnung „typisiert und überwacht vom Staatlichen Materialprüfungsamt Berlin-Dahlem"

[1]) Kunstharz-P r e ß m a s s e n im Sinne dieser Bekanntmachung sind härtbare, warm zu verpressende Erzeugnisse (auf der Grundlage von Phenol- oder Harnstoff-Harzen) sowie nicht härtbare Spritzgußmassen (Typ A), welche als ungeformte Halbfabrikate in den Handel gebracht werden.
Kunstharz-P r e ß s t o f f e sind die aus Kunstharz-Preßmassen im Preß- oder Preßspritz-Verfahren hergestellten Formkörper (Preßlinge); vgl. auch Begriffs-Erklärungen im VDE 0320/1939 „Leitsätze für nichtkeramische gummifreie Isolierstoffe".
[2]) Vgl. folgende „Bekanntmachung über typisierte und überwachte Preßstoffe", Tafel I und III.

in den Handel gebracht werden. Die Verpackungen dieser Preßmassen tragen das Ueberwachungszeichen ⊕ (vgl. DIN 7702), in welchem das Firmen-Kennzeichen und das Typzeichen angegeben sind.

Pressereien, die nicht selber Kunstharz-Preßmassen herstellen, dürfen für die Herstellung von Kunstharz-Preßteilen (Kunstharz-Preßstoffen) mit dem Ueberwachungszeichen nur die in den Tafeln A, B und C aufgeführten Kunstharz-Preßmassen verwenden.

Diejenigen Kunstharz-Preßmassen, welche als Preßstoff den in der Typisierung[3]) aufgeführten elektrischen Sonderanforderungen genügen und als Zusatz zum Typzeichen einen Stern führen, sind in Tafel C gesondert aufgeführt.

Die in den Zeitschriften: Kunststoffe Bd. 29 (1939) S. 93, Kunststoff-Technik und Kunststoff-Anwendung Bd. 9 (1939) S. 96 und ETZ Bd. 60 (1939) S. 347 bekanntgegebenen Tafeln A, B und C sind hiermit ungültig.

Berlin-Dahlem, den 22. 2. 1940.

Der Präsident:
i. V. Kindscher

[3]) Vgl. „Typisierung der gummifreien nichtkeramischen Isolierpreßstoffe", Kunststoffe Bd. 27 (1937) S. 330; Plastische Massen Bd. 7 (1937) S. 339; ETZ Bd. 58 (1937) S. 1254

Tafel A

Firma	Firmen-Kennzeichen im Ueberwachungszeichen	Typ 11		Typ 12		Typ M		Typ 0		Typ S		Typ T1	
		Handelsbezeichnung	Reihen-[1])	Handelsbezeichnung	Reihen-[1])	Handelsbezeichnung	Reihen-[1])	Handelsbezeichnung	Reihen-[1])	Handelsbezeichnung	Reihen-[1])	Handelsbezeichnung	Reihen-[1])
Dr. Kurt Albert G. m. b. H., Chemische Fabriken, Amöneburg bei Wiesbaden-Biebrich	Al	Albert-Preßmasse Typ 11	9200	Albert-Preßmasse Typ 12	9000	—		—		Albert-Schnellpreßmasse Typ S	200 400 500 800 900 1300 2000 2500 3000 3500 4000 2200 2700 3700	Albert-Preßmasse Typ T1	6500
Bakelite Gesellschaft m. b. H., Erkner bei Berlin	Ba	Bakelite-Schnellpreßmasse Typ 11	1 100 G	Bakelite-Schnellpreßmasse Typ 12	1 200 A	Bakelite-Schnellpreßmasse Typ M	M 1300	Bakelite-Schnellpreßmasse Typ 0	0 300	Bakelite-Schnellpreßmasse Typ S	S 600 S 700 S 800 S 900 S 5600 S 6600	Bakelite-Schnellpreßmasse Typ T1	T 1400
Dynamit-Actien-Gesellschaft vormals Alfred Nobel & Co., Abteilung Celluloid- und Kunststoff-Fabrik, Werk Troisdorf (Bez. Köln); Verkauf durch Venditor, Kunststoff-Verkaufsges. m. b. H., Troisdorf (Bez. Köln)	DAG	Trolitan I/37	7000	Trolitan I Trolitan AF	7000 7000	Trolitan AW	7000	—		Trolitan S Trolitan SP-R	2000 2100 2200 2300 2400 2600 2700 2500 2800 2900	Trolitan L Trolitan CL	7000 7000
Internationale Galalith-Gesellschaft Hoff & Co., Hamburg-Harburg 1	IGG	—		—		—		—		Kerit-Schnellpreßmasse Typ S	0 100 300 1000 1100 1200 2000 11000 11500 12000	—	
New-York Hamburger Gummi-Waaren Compagnie, Abt. Kunststoffe, Hamburg 33	NYH	Faturan-Gesteinsmehlmasse Typ 11	700 700 AF	Faturan-Asbestfasermasse Typ 12	900	—		—		Faturan-Schnellpreßmasse Typ S	0 100 500 2000 2500 500 AF	Faturan-Textilfasermasse Typ T1	600 600 AF
Aug. Nowack Aktiengesellschaft, Bautzen	No	Neoresit-Preßmasse Typ 11	400 4000	Neoresit-Preßmasse Typ 12	500	Neoresit-Asbestschnurpreßmasse Typ M	600	Neoresit-Preßmasse Typ 0	0	Neoresit-Preßmasse Typ S	100 200 1000 2000 5000 6000	Neoresit-Textilpreßmasse Typ T1	7000
Dr. F. Raschig G. m. b. H., Chemische Fabrik, Ludwigshafen a. Rhein	Ra	Resinol Typ 11	400	—		—		—		Resinol Typ S	100 200 300 700 800 900	—	
										Resinol Typ S ammoniakfrei	10100 10200 10700 10800		
H. Römmler Aktiengesellschaft, Spremberg (Nd.-Lausitz)	HRS	Hares F 1	1930 ...1949	Hares F 2	1950 ...1969	Ralotext	1910 ...1929	—		Hares S	H O 0 700 900 1700 2000	Hares L 1	2000 ...2049
Vereinigte Chemische Fabriken Kreidl, Heller & Co. Nfg., Wien, XXI	VCF	—		—		—		—		Deligna S	100 300 1000 4000 6000	—	

[1]) Die Reihenbezeichnung 0 umfaßt die Nummern 0 bis 99, die Reihenbezeichnung 100 umfaßt die Nummern 100 bis 199 usw.

Appendix III: Abbreviations and Company Initials

The Tables A, B and C announced in the periodicals: Plastics Vol. 29 (1939) p. 93, Plastic Technology and Plastic Application Vol. 9 (1939) p. 96, and ETZ Vol. 60 (1939) p. 347 are herewith invalid.

Berlin-Dahlem, 22 Feb 1940.

The President:
by authority: K i n d s c h e r"

Firm	Table A [Tables B, C; see also *Appendix IV*]
[AE]	(unconfirmed)
[Agfe]	(unconfirmed)
Al	Dr. Kurt Albert G.m.b.H., Chemische Fabriken, Amöneburg bei Wiesbaden-Biebrich
Ba	Bakelite Gesellschaft G.m.b.H, Erkner bei Berlin
[Bi]	Bisterfeld & Stolting, Inhaber: Ernst Bisterfeld, Radevormwald (Rhld.) [70]
[BJB]	Brökelmann, Jaeger & Busse, Neheim i. Westf. [also L7]
DAG	Dynamit-Actien-Gesellschaft (formerly Alfred Nobel & Co., Abteilung Celluloid-und Kunstoff-Fabrik, Werk Troisdorf Bez. Köln, and marketed and sold through Venditor, Kunstoff-Verkaufsges. G.m.b.H., Troisdorf Bez. Köln [43]
[DS]	(unconfirmed)
[DSK]	Reppel & Vollmann, Kierspe [dsk]
[E]	(unconfirmed)
[ET]	(unconfirmed)
[F.M.]	Fr. Möller, Brackwede i. Westf. [V5]
[F&S]	(unconfirmed)
[FW]	(unconfirmed)
[G]	(unconfirmed)

[GMG]	Gebrüder Merten, Gummersbach [also 24]
[G.S.]	Gebr. Spindler Betr.-Kom.-Ges., KG, Köppelsdorf i. Thür. [E7]
HRS	H. Römmler Aktiengesellschaft, Spremberg Nd.-Lausitz [32]
[H.S.]	Elektrotechnische Metallwarenfabrik Storch und Stehmann G.m.b.H., Ruhla i. Thür.[M8]
I.G.G.	Internationale Galalith-Gesellschaft Hoff & Co., Hamburg-Harburg [Z1]
[JMB]	Jos. Mellert, Bretten i. Baden [M9]
[JML]	(unconfirmed)
[KA]	(unconfirmed)
[KAW]	(unconfirmed)
[KB]	Kunstharz-Presserei Schwaben, Ingenieur Otto Single, Plochingen i. Württ. [W0]
[KB]	Heinrich Knöll, Gross-Bieberau i. Odenwald [W8]
[KH]	(unconfirmed)
[KLCO]	(unconfirmed)
[KP]	Karl Potthof, Presswerk, Solingen-Ohlings [N9]
No	Aug. Nowack Aktiengesellschaft, Bautzen
NYH	New York-Hamburger Gummi-Waaren Compagnie, Abt. Kunststoffe, Hamburg
P-A.G.	Preßwerk A.G., Essen [45]
[PKC]	Paul Kuhbier & Co., Wipperfürth (Rhld.) [Z4]
[PS]	Pressstoffwerk Schöppenstedt, Paul Schnake, Schöppenstedt [81])
Ra	Dr. F. Raschig G.m.b.H., Chemische Fabrik, Ludwigshafen a. Rhein

Appendix III: Abbreviations and Company Initials 127

[RG]	(unconfirmed)
[Rö]	Pressmaterial-Werk Hermann Römmler und Schumann K.G., Berlin-Friedenau [74]
[RWS]	Rheinische-Westfalische Sprengstoff Actiengesellschaft, Werk Stadeln, bei Fürth [dnf], (acquired later by Alfred Nobel und Co, then in turn even later by Dynamit A.G., [43])
[Sch.]	Apparatebauanstalt Schneider & Co. Breslau-Gr. Ohlewiesen [U4]
[S+M]	(unconfirmed)
[t]	another mark used by Dynamit AG, Werk Troisdorf, Troisdorf [43]
[trf]	another mark used by Dynamit AG, Werk Troisdorf, Troisdorf [43]
[TP]	another mark used by Dynamit AG, Werk Troisdorf, Troisdorf [43]
VCF	Vereinigte Chemische Fabriken Kreidl, Heller & Co., Nfg., Wein

Appendix IV:
Alphabetical Listing of *Kunststoff* Manufacturers

The second part of the 1940 article *Bekanntmachung über Kunstharz-Preßmassen für typisierte und überwachte Preßstoffe* (Announcement Concerning Synthetic Resin Molding Compounds for Standardized and Inspected Moldings) as it appeared in *Kunststoffe* Bd.30, Heft 3, pgs. 82-88, is transcribed, translated (by Paul Seibold) and provided for the reader below, primarily as an *alphabetic* listing of the wartime German plastics manufacturers. The original document was published with tables listing the predominant trade names and material types for each firm, which is not provided here. Notations in brackets are cross-references to known MPD codes as tabulated by the author.

"State Material Testing Office in Berlin-Dahlem
Unter den Eichen 87
Telephone: 76 27 51

Announcement Concerning Standardized and Inspected Moldings [60]

The moldings inspected on the basis of the "Standardizing of Rubber-free Non ceramic Insulating Moldings" [61] by the State Material Testing Office Berlin-Dahlem according to agreement with the "Technical Alliance of Manufacturers of Standardized Molding Compounds and Moldings (Registered Association)," Berlin, are compiled in the following Tables I, II and III.
    Those types of moldings which in addition to the normal mechanical, thermal
and electrical conditions satisfy also the special electrical requirements listed in the standardization and bear a star in addition to the type symbol are specially listed as Table III.
    The Tables I, II and III announced in Plastics Vol. 29 (1939) p. 95 f., Plastic Technology and Plastic Application Vol. 9 (1939) p. 99 f. and ETZ Vol. 60 (1939) p.
377 f. are herewith invalid.

Berlin-Dahlem, 22 Feb 1940

the President:
by authority: K i n d s c h e r"

Appendix IV: Alphabetical Listing of Kunststoff Manufacturers

82   Umschau aus Schrifttum und Technik   Kunststoffe Bd. 30 1940 Heft 3

Staatliches Materialprüfungsamt in Berlin-Dahlem
Unter den Eichen 87
Fernsprecher: 76 27 51

Bekanntmachung über typisierte und überwachte Preßstoffe[1]).

Die auf Grund der „Typisierung der gummifreien nichtkeramischen Isolierpreßstoffe"[2]) vom Staatlichen Materialprüfungsamt

[1]) Preßstoffe im Sinne dieser Bekanntmachung sind die in der „Typisierung der gummifreien nichtkeramischen Isolierpreßstoffe" (vgl. Anm. 2) aufgeführten Stoffe, welche aus Preß- oder Spritzmassen im Preß-, Preßspritz- oder Spritzverfahren als Formkörper (Preßlinge) hergestellt werden.

[2]) Vgl. Kunststoffe Bd. 27 (1937) S. 330; Plastische Massen Bd. 7 (1937) S. 339; ETZ 58 (1937) S. 1254.

Berlin-Dahlem gemäß Vertrag mit der „Technische Vereinigung der Hersteller typisierter Preßmassen und Preßstoffe e. V.", Berlin, überwachten Preßstoffe[1]) sind in den folgenden Tafeln I, II und III zusammengestellt.

Diejenigen Preßstoff-Typen, welche außer den normalen mechanischen, thermischen und elektrischen Bedingungen noch den in der Typisierung aufgeführten elektrischen Sonderanforderungen genügen und als Zusatz zum Typzeichen einen Stern führen, sind als Tafel III gesondert aufgeführt.

Die in den Zeitschriften: Kunststoffe Bd. 29 (1939) S. 95 f., Kunststoff-Technik und Kunststoff-Anwendung Bd. 9 (1939) S. 99f. und ETZ Bd. 60 (1939) S. 377f. bekanntgegebenen Tafeln I, II und III sind hiermit ungültig.

Berlin-Dahlem, den 22. 2. 1940

Der Präsident:
i. V. Kindscher.

Tafel I

Firma	Firmen-Kennzeichen im Überwachungszeichen	11	12	M	0	S	T1	T2	T3	Z1	Z2	Z3	K
Acla, Rheinische Maschinen-, Leder- u. Riemenfabrik Aktiengesellschaft, Köln-Mülheim	0 Z	—	—	—	—	Preßstoff Typ S	—	—	—	—	—	—	—
Emil Adolff, Abt. Kunstharz-Presswerk, Reutlingen i. Württ.	H 2	—	—	—	—	Preßstoff Typ S	—	—	—	—	—	—	—
Gebrüder Adt Aktiengesellschaft, Wächtersbach H.-N.	36	—	siehe Tafel II				—						
Agalitwerk Milspe, Kattwinkel & Co., Milspe i. Westf.	T 8	—	—	—	—	Preßstoff Typ S	—	—	—	—	—	—	—
Allgemeine Elektricitäts-Gesellschaft, Fabriken Hennigsdorf, Hennigsdorf (Osthavelland)	38	Tenacit Typ 11	Tenacit Typ 12	Tenacit Typ M	Tenacit Typ 0	Tenacit Typ S	Tenacit Typ T1	Tenacit Typ T2	Tenacit Typ T3	Tenacit Typ Z1	Tenacit Typ Z2	Tenacit Typ Z3	Tenacit Typ K
Allgemeine Elektricitäts-Gesellschaft, Fabriken Annaberg, Annaberg i. Erzgeb. Preßwerk Scheibenberg i. Erzgeb.	38 A	„	„	„	„	„	„	„	„	„	„	„	„
Alusil-Preßstoffwerk Eugen Gassmann, Probstzella i. Thür.	F 1	—	Alusil 12	—	—	Alusil S	—	—	—	—	—	—	—
Robert Anke, elektrot. Porzellanfabrik & Kunstharzpresserei, Oelsnitz i. Vogtl.	F 4	—	—	—	—	Preßstoff Typ S	—	—	—	—	—	—	Preßstoff Typ K
Apparatebauanstalt Schneider & Co., Breslau-Gr. Ohlewiesen	U 4	—	—	—	—	Escolith S	—	—	—	—	—	—	—
Bachmann & Leichsenring, Berlin-Neukölln	V 4	—	—	—	—	Preßstoff Typ S	—	Preßstoff Typ T2	—	—	—	—	—
Ernst Backhaus & Co., Kierspe-Bahnhof i. Westf.	53	—	—	Preßstoff Typ M	—	Preßstoff Typ S	—	Preßstoff Typ T2	—	—	—	Preßstoff Typ Z3	Preßstoff Typ K
Otto Backhaus, Bollwerk i. Westf.	F 0	Preßstoff Typ 11	—	—	—	Preßstoff Typ S	—	—	—	—	—	—	Preßstoff Typ K
Bamberger Industrie-Gesellschaft, Bamberg i. Bayern	T 4	—	—	—	—	Preßstoff Typ S	—	—	—	—	—	—	—
Barth, Klemm & Co., Leipzig W 31	0 T	—	—	—	—	—	—	Preßstoff Typ T2	—	—	—	—	—
Bayerische Elektrozubehör G.m.b.H., Lauf bei Nürnberg	67	—	Bezeg 12	—	—	Bezeg S	Bezeg T1	Bezeg T2	—	—	—	—	Bezeg K
Bebrit-Preßstoffwerke G.m.b.H., Bebra u. C. & F. Schlothauer G.m.b.H., Ruhla (Thür.), Bebra (H.-N.)	22	Debrit 11	—	—	—	Debrit S	—	Preßstoff Typ T2	—	Preßstoff Typ Z1	Preßstoff Typ Z2	—	Preßstoff Typ K
Bender & Wirth, Kierspe-Bahnhof i. Westf.	M 1	—	—	—	—	Preßstoff Typ S	—	—	—	—	—	—	—
Bergfeld & Heider, Burscheid (Bez. Düsseldorf)	2 X	—	—	—	—	Preßstoff Typ S	—	—	—	—	—	—	—
Gebr. Berker, Schalksmühle i. Westf.	80	—	—	—	—	Isolierpanzer Typ S	—	Isolierpanzer Typ T2	—	—	Isolierpanzer Typ Z2	—	Isolierpanzer Typ K
Porzellanfabrik Bernhardshütte G.m.b.H., Blechhammer bei Sonneberg i. Thür.	93	Preßstoff Typ 11	Preßstoff Typ 12	Preßstoff Typ M	—	Preßstoff Typ S	—	Preßstoff Typ T2	—	Preßstoff Typ Z1	Preßstoff Typ Z2	—	Preßstoff Typ K
Bezet-Werk Hermann Buchholz, Motzen (Krs. Teltow)	76	—	Preßstoff Typ 12	—	Preßstoff Typ 0	Preßstoff Typ S	Preßstoff Typ T1	Preßstoff Typ T2	—	—	Preßstoff Typ Z2	Preßstoff Typ Z3	—
Bisterfeld & Stolting, Inhaber: Ernst Bisterfeld, Radevormwald (Rhld.)	70	—	Werkstoff Typ 12	Werkstoff Typ M	Werkstoff Typ 0	Werkstoff Typ S	Werkstoff Typ T1	Werkstoff Typ T2	—	Werkstoff Typ Z1	Werkstoff Typ Z2	—	Werkstoff Typ K
Blumberg & Co., gegr. 1885, Lintorf (Bez. Düsseldorf)	W 6	—	—	—	—	Preßstoff Typ S	—	Preßstoff Typ T2	—	—	—	—	—
H. Bodenmüller, Ing., Stuttgart-Zuffenhausen	L 8	—	—	—	—	Preßstoff Typ S	—	—	—	—	—	—	—
Böhmische Kontaktwerke Aktiengesellschaft, Komotau (Reichsgau Sudetenland)	1 N	—	—	—	—	Preßstoff Typ S	—	—	—	—	—	—	—
Bonner Keramik Aktiengesellschaft, Bonn a. Rhein	W 2	—	—	—	—	Preßstoff Typ S	—	Preßstoff Typ T2	—	—	—	—	—
Robert Bosch G.m.b.H., Metallwerk, Stuttgart-Feuerbach	55	—	—	—	Resiform 0	Resiform S	Resiform T1	Resiform T2	—	—	—	—	—
Max Braun, Frankfurt a. Main	3 A	—	—	—	—	Preßstoff Typ S	—	—	—	—	—	—	—
Ernst Bremicker, Ing., Kierspe-Bahnhof i. Westf.	61	—	—	—	—	Preßstoff Typ S	—	—	—	—	—	—	Preßstoff Typ K
Gebr. Broghammer, Schramberg (Schwarzwald)	F 8	—	—	—	—	Preßstoff Typ S	—	—	—	—	—	—	—

Kunststoffe

Firm   Table [s] I [II, III]

[-A-]

Acla, Rheinische Maschinen-, Leder- u. Riemenfabrik, Aktiengesellschaft,
  Köln-Mülheim [0Z]
Emil Adolff, Abt. Kunstharz-Presswerk, Reutlingen i. Württ. [H2]
Gebrüder Adt Aktiengesellschaft, Wächtersbach [36]
Agalitwerk Milspe, Kattwinkel & Co., Milspe i. Westf. [T8]
[Dr. Kurt Albert G.m.b.H., Chemische Fabriken, Amöneburg bei
  Wiesbaden-Biebrich [Al]]
Allgemeine Elektricitäts-Gesellschaft, Fabriken Hennigsdorf,
  Henningsdorf (Osthavelland) [38]
Allgemeine Elektricitäts-Gesellschaft, Fabriken Annaberg, Annaberg i. Erzgeb.,
  Presswerk Scheibenberg i. Erzgeb. [38A]
Alusil-Preßstoffwerk Eugen Gassmann, Probstzella i. Thür. [F1]
Robert Anke, elektrot. Porzellanfabrik & Kunstharzpresserei, Oelsnitz i. Vogtl. [F4]
Apparatebauanstalt Schneider & Co. Breslau-Gr. Ohlewiesen [U4]

[-B-]

Bachmann & Leichsenring, Berlin-Neuköln [V4]
Ernst Backhaus & Co., Kierspe-Bahnhof i. Westf. [53]
Otto Backhaus, Bollwerk i. Westf. [F0]
[Bakelite Gesellschaft G.m.b.H, Erkner bei Berlin [Ba]]
Bamberger Industrie-Gesellschaft, Bamberg, i. Bäyern [T4]
Barth, Klemm & Co., Leipzig [0T]
Bayerische Elektrozubehör G.m.b.H., Lauf bei Nürnberg [67]
Bebrit-Preßstoffwerke G.m.b.H., Bebra u. C. & F, Schlothauer G.m.b.H., Ruhla (Thür.);
  Bebra [22]
Bender & Wirth, Kierspe-Bahnhoff i. Westf. [M1]
Bergfeld & Heider, Burscheid (Bez. Düsseldorf) [2X]
Gebr. Berker, Schalksmühle i. Westf. [80]
Porzellanfabrik Bernhardshütte G.m.b.H., Blechhammer bei Sonneberg i. Thür. [93]
Bezet-Werk Hermann Buchholz, Motzen (Krs. Teltow) [76]
Bisterfeld & Stolting, Inhaber: Ernst Bisterfeld, Radevormwald (Rhld.) [70]
Blumberg & Co., Lintorf (Bez. Düsseldorf) [W6]
H. Bodenmüller, Ing., Stuttgart-Zuffenhausen [L8]
Böhmische Kontactwerke Aktiengesellschaft, Komotau (Reichsgau Sudetenland) [1N]
Bonner Keramik Aktiengesellschaft, Bonn a. Rhein [W2]
Robert Bosch G.m.b.H., Metallwerk, Stuttgart-Feuerbach [55]
Max Braun, Frankfurt a. Main [3A]
Ernst Bremicker, Ing., Kierspe-Bahnhof i. Westf. [61]
Gebr. Broghammer, Schramberg (Schwarzwald) [F8]
Brökelmann, Jäger & Busse, Neheim i. Westf. [L7]

Appendix IV: Alphabetical Listing of Kunststoff Manufacturers

Richard Brünner, Elektrotechnische Fabrik, Wien, VII. [N8]
Brunnquell & Co., Sonderhausen i. Thür. [2H]
Karl Buchruker, München 58 [U5]
Wilh. Burgbacher K.G., Neukirch i. Baden (Station Furtwangen) [T2]
Busch-Jaeger Lüdenscheider Metallwerke Aktiengesellschaft, Lüdenschied i. Westf. [23]

[-C-]

[-D-]

Gebr. Dahlhaus, Schalksmühle i. Westf. [Z6]
Dr. Deisting & Co., Gesellschaft mit beschrankter Haftung, Kierspe i. Westf. [39]
Deutsche Legrit- Ges. m.b.H. Berlin [A4]
Deutsche Philips-Gesellschaft m.b.H., Berlin W62 [58]
Dornseif & Linde, Kierspe i. Westf. [0M]
Friedrich Dörscheln, Lüdenscheid i. Westf. [99]
Dralowid-Werk der Steatit-Magnesia-Aktiengesellschaft, Teltow bei Berlin [W3]
Dynamit-Actien-Gesellschaft vormals Alfred Nobel & Co., Abteilung Celluloid-und-
    Kunststoff-Fabrik, Werk Troisdorf (Bez. Köln); verkauf durch Venditor, Kunststoff-
    Verkaufsges. m.b.H, Troisdorf (Bez. Köln) [43]

[-E-]

Eisele Auto-Electric, Inhaber Karl Klos, Frankfurt a. Main [Z5]
[Metallwerk Elektra G.m.b.H., Gummersbach (Rhld.) [U6]]
Elektrotechnische Fabrik J. Carl, Gesellsch. m. beschr. Haftung,
    Oberweimar i. Thür. [E0]
Elektrotechnische Fabrik Weber & Co., Komm.-Ges., Kranichfeld i. Thür. [98]
Ellinger & Geißler. Dorfhain (Bez. Dresden) [54]
Erlemann & Co., Bergerhof (Rhld.) [0H]

[-F-]

"Feba" Fabr. Elektr. Bedarfsartikel Stückrath K.-G., Berlin-Köpenick [U1]
Josef Feix Söhne, Gablonz a. Neiße (Reichsgau Sudetenland) [2E]
Fischer & Klüppelberg vorm. W. Dörner & Co., G.m.b.H., Radevormwald (Rhld.) [Y7]
Frankl & Kirschner, Elektrizitätsgesellschaft, Mannheim-Neckarau [M7]
Fresen & Co., Lüdenscheid i. Westf. [V6]
Robert Friedrich, Abteilung III [Kartonagenfabrik], Annaberg [-Buchholz] i. Erzgeb. [2N]
Friemann & Wolf G.m.b.H., Zwickau i. Sa. [H8]
Futurit-Werk Aktiengesellschaft, Wien [92]

132　　　　　　　　　　　Kunststoffe

[-G-]

Oskar Gaudlitz, Coburg [60]
Wilhelm Geiger G.m.b.H, Lüdenscheid i. Westf. [71]
Gerdes & Co., Schwelm i. Westf. [H4]
Kunstharz-Presserei Carl Germer, Berlin NW 87 [M3]
Christian Geyer, Nürnberg [83]
Richard Giersiepen, Bergisch Born (Rhein-Wupperkreis) [M5]
Goltzsche, Merlin & Sohn, Großröhrsdorf i. Sa. [V3]
Ernst Gomolka, Zehdenick (Mark) [X8]
[Julius Karl Görler, Transformatorenfabrik, Berlin-Charlottenburg 1 [X6]]
Goseberg & Grashoff, Kierspe-Bahnhof i. Wesf. [Y3]
Ernst Gösser, Iserlohn i. Westf. [X9]
Graewe und Co., Menden i. Westf. (Kreis Iserlohn) [E8]
Grelit Kunstharz-Presswerk Grohmann, Pietschmann & Co.,
　　Nixdorf (Reichsgau Sudetenland) [1Y]

[-H-]

Kunstharz-Presswerk der Hanf-, Jute- u. Textil-Industrie A.G., Wien I [0L]
Heliowatt-Werke, Elektrizitäts-Aktiengesellschaft, Berlin-Charlottenburg 4 [35]
Hering & Co., Iserlohnerheide bei Iserlohn, Post über Schwerte/Ruhr [1X]
Fritz Heublein, Neustadt bei Coburg [T7]
Hans Heußinger, Ing., Kunstoff-Presserei, Nürnberg [1T]
Kunstharzpresserei M. Hildebrandt und E. Hammerschmidt, Brand-Erbisdorf i. Sa. [N5]
Richard Hirschmann, Esslingen a. N. [2M]
Paul Hochköpper und Co., Lüdenscheid i.Westf. [87]
Metallwerke Adolf Hopf, Aktiengesellschaft, Tambach-Dietharz i. Thür. [0E]
Hoppmann und Mulsow, Hamburg [L1]
Herbert Horn, Pulsnitz i. Sa. [Y0]
Gebrüder von der Horst, Lüdenscheid i. Westf. [H7]

[-I-]

Internationale Galalith-Gesellschaft Hoff & Co., Hamburg-Harburg [Z1]
Isola Werke A.G., Birkesdorf -Düren (Rhld.) [40]
Isopress-Werk G.m.b.H., Berlin-Oberschöneweide [L2]
Dr. Paul Isphording, Kunstharzpresserei, Wepritz bei Landsberg (Warthe) [0N]

[-J-]

Erich Jaeger K.-G., Bad Homburg v.d. Höhe [95]
Paul Jatow, Dodendorf (Bez. Magdeburg) [E4]
Paul Jordan, Elekrotechnische Fabrik, Berlin-Steglitz [N4]

[-K-]

Gebrüder Kaiser & Co., Neheim-Ruhr [1M]
Kaiser und Spelsberg, Schalksmühle i. Westf. [M6]
Robert Karst, Berlin [L9]
Franz Kirsten, Elektrotechische Spezialfabrik, Bingerbrück a. Rhein [2F]
Willi Kleinau, Kunstharz-Preß- und Spritzwerk, Berlin O 112 [3H]
Ernst Klever, Kunstharz-Presserei, Solingen-Nord [2Y]
Heinrich Knöll, Gross-Bieberau i. Odenwald [W8]
Köditz & Co., Metallwarenfabrik u. Presserei, Langewiesen i. Thür. [Z9]
Heinrich Kopp G.m.b.H., Sonneberg i. Thür. [W1]
Leopold Kostal, Lüdenscheid i. Westf. [72]
Signalapparatefabrik Julius Kräcker Aktiengesellschaft, Berlin SW 61 [V8]
Theodor Krägeloh & Comp., Dahlerbrück i. Westf. [85]
[Vereinigte Chemische Fabriken Kreidl, Heller & Co., Nfg., Wein [VCF]]
Hugo Krieger und [Hermann] Faudt, Berlin SW 68 [78]
Kronacher Porzellanfabrik Stockhardt und Schmidt-Eckert, Kronach i. Bayern [M4]
Krone & Co, Berlin-Baumschulenweg [L6]
Kugella vormals Max Roth G.m.b.H., Mittelschmalkalden (Post Wernshausen) [X7]
Paul Kuhbier & Co., Wipperfürth (Rhld.) [Z4]
Kunstharz-Presserei Schwaben, Ingenieur Otto Single, Plochingen i. Württ. [W0]

[-L-]

Otto Langmann, Kunstharz-Presswerk, Hagen i. Westf. [H3]
Linden & Co. K.G., Lüdenscheid i. Westf. [79]
Lindner & Co., Jecha-Sondershausen i.Thür. [A8]
Gerhard Lischke, Kunstharzpresserei, Oppach i. Sa. [2U]
[Heinrich List Elektro-Meßgerätebau, Berlin-Steglitz [4Z]]
Lohmann und Welschehold, Meinerzhagen i. Westf. [E9]
C. Lorenz Aktiengesellschaft, Berlin-Tempelhof [Y9]
Carl Friedr. Lübold, Lüdenscheid i. Westf. [M0]

[-M-]

Märkische Elektro-Industrie Adolf Vedder K.G., Schalksmühle i. Westf. [N2]
Meirowsky und Co. Aktiengesellschaft, Porz a. Rhein [L4]
Jos. Mellert, Bretten i. Baden [M9]
Radio H. Mende & Co., Abt. Presswerk "Mendelith", Dresden [56]

H. Merlet, Kunstoffpresswerk, Neudorf b. Glabonz a. Neiße
   (Reichsgau Sudetenland) [3T]
Gebrüder Merten, Gummersbach (Rhld.) [24]
Metallwerk Elektra G.m.b.H., Gummersbach (Rhld.) [U6]
Mix & Genest Aktiengesellschaft, Berlin-Schöneberg [68]
Fr. Möller, Brackwede i. Westf. [V5]
Karl Müller, Solingen-Merscheid [1A]

[-N-]

Otto Nettelbeck, Berlin O 17 [V2]
New-York Hamburger Gummi-Waaren Compagnie, Hamburg [21]
[Aug. Nowack Aktiengesellshaft, Bautzen [No]]

[-O-]

Ostland [-Werke] G.m.b.H., Königsberg i. Pr. [X5]
Ostpreussisches Kunstharz-Presswerk, Gedenk, Blank & Naujoks-Erben,
   Königsberg i. Pr [1V]

[-P-]

Ing. Fr. August Pfeffer, [Sonneberg-]Oberlind i. Thür. [0Y]
Carl Pfestorf, Abt. Kunstharzpresserei, Tambach-Dietharz i. Thür. [2A]
Phenoplast, Bischoff & Co., Kom.-Ges., Eberswalde [-Finow] [V1]
Plate & Voerster, Kierspe i. Westf. [Z0]
Porzellanfabrik Theodor Pohl, Schatzlar i. Riesengebirge [L5]
C. Pose, Wehrausrüstungen, Berlin O 34 [0F]
Julius Posselt [Bakelite], Gablonz a. Neiße (Reichsgau Sudetenland) [1W]
Kurt Postel, Köln, Höhenberg [N6]
Karl Potthoff, Presswerk, Solingen-Ohligs [N9]
Preh, Elektrofeinmechanische Werke, Bad Neustadt/Saale [U2]
Pressmaterial-Werk Hermann Römmler und Schumann K.G., Berlin-Friedenau [74]
Presstoffwerk Nürnberg, Gebrüder Klein, Nürnberg [V9]
Pressstoffwerk Schöppenstedt, Paul Schnake, Schöppenstedt [81]
Presswerk A.G., Essen [45]
Presswerk Königstein, R. Saring, Königstein, Sächs. Schweiz [65]
Presswerk Mollberg & Co., Hofgeismar, Bez. Kassel [0V]
Presswerk Kurt Wentzel, Inhaber Anton Jacob, Berlin-Steglitz [X3]
Presswerk Winkel, Schulte & Conze, Herscheid i. Westf. [51]

[-Q-]

Wilhelm Quante, Spezialfabrik für Apparate der Fernmeldetechnik, Inhaber:
Hermann Quante, Wuppertal-Elberfeld [73]

[-R-]

[Dr. F. Raschig G.m.b.H., Chemische Fabrik, Ludwigshafen a. Rhein [Ra]]
Anton Reiche A.-G., Dreden [3M]
Reicolit-Presswerk, Cuno Heinzelmann-Hasberg, Berlin C 2 [N7]
Gebrüder Reiher K.G., vormals Aktiengesellschaft für Elektrotechnik,
  Braunschweig [L0]
Reininghaus & Co., Beleuchtungskörper-Fabrik, Lüdenscheid i. Westf. [3F]
Rheinisch-Westfälisches Kunstoffwerk G.m.b.H., Mülheim-Ruhr [Z8]
Ludwig Richter, elektrotechn. Spezialfabrik, Görlitz [3Y]
Richard Rinker G.m.b.H., Menden (Krs. Iserlohn) [X0]
H. Römmler Aktiengesellschaft, Spremberg (Nd.-Lausitz) [32]
[Pressmaterial-Werk Hermann Römmler und Schumann K.G., Berlin-Friedenau [74]]
Hermann Ros, Coburg [66]
G. Rövenstrunk & Co., Elspe bei Brügge i. Westf. [X4]
George Rüger & Co., vorm. Elektra G.m.b.H., Elektrotechnische Fabrik,
  Essen-Heidhausen [3X]
Adolf Ruoff, Kunstoff-Presswerk, Radevormwald (Rhld.) [Y8]

[-S-]

Elektro-Glimmer-und Preßwerke Scherb & Schwer K.-G. vorm.
  Jareslaw, Berlin-Weißensee [T6]
Schieck-Instrument Wilhelm Wolkersdorf, Berlin [1E]
Schmachtenberg & Türck, Solingen-Wald [Y5]
Heinrich Schmidberger, Wien [Z7]
Schmidt & Co. K.G., Metallwarenfabrik, Schweim i. Westf. [3E]
[Apparatebauanstalt Schneider & Co. Breslau-Gr. Ohlewiesen [U4]]
Schriever & Co., Kunstoffpresserei, Kierspe-bahnhof i. Westf. [2W]
Ludwig Schröder, Schalksmühle i. Westf. [U9]
Ferdinand Schuchhardt, Berliner Fernsprch-und Telegraphenwerk A.G.,
  Berlin SO 16 [Y1]
F.F.A. Schulze, Metallwarenfabrik, Berlin [1H]
Max Schulze, Meißen i. Sa. [F5]
Fabrik isolierter Drähte u. Schnüre, Schulze, Schneider & Dort G.m.b.H.,
  Schönow, Post Bernau bei Berlin [H9]
Seckelmann & Co., Lüdenscheid i. Westf. [59]
Karl Friedr. Seiter K.-G., Bollwerk i. Westf., Post Oberbrügge [0X]
Ernst Albert Senf, Kunstharzpresserei, Bautzen i. Sa. [W7]
Gebr. Sieling, Metallwaren-Fabrik und Kunstharz-Presswerk, Lüdenscheid i. Westf. [3L]

Siemens-Schuckert-Werke, Aktiengesellschaft, Abteilung Isolierstoffe (SK 4), Berlin-Siemensstadt (Gartenfeld) [34]
Siemens-Schuckert-Werke, Aktiengesellschaft, Wiener Maschinen- und Apparate-Werk, Wien [M2]
Wilhelm Sihn jr., Niefern i. Baden [U7]
Singer Nähmaschinen Aktiengesellschaft, Fabrik Wittenberge, Bez. Potsdam [1F]
Gebr. Spindler Betr.-Kom.-Ges., KG, Köppelsdorf i. Thür. [E7]
Starkstrom-Apparatebau G.m.b.H., Berlin; Presserei: Zweigwerk Buschullersdorf bei Reichenberg (Reichsgau Sudentenland) [2T]
Franz Stauch, Presswerk, Unterrodach i. Ofr. [F2]
Kurt Steidel, Berlin [86]
W. Stiefeling, Berlin [X1]
Elektrotechnische Metallwarenfabrik Storch & Stehmann G.m.b.H., Ruhla i. Thür. [M8]
Strauss & Co., Schmölln i. Thür., Abteilung Kunstharzpresserei, Auma i. Thür. [W5]
Süddeutsche Isolatoren-Werke G.m.b.H, Freiburg im Breisgau [25]
Röhrenwerk Johannes Sturmann G.m.b.H., Arnsberg i. West. [2Z]
Sursum Elektr.-Gesellschaft Leyhausen & Co., Nürnberg [N1]

[-T-]

Paul Teich, Berlin [82]
Telefonbau und Normalzeit G.m.b.H., Frankfurt a. Main [3N]
Thega-Kontakt G.m.b.H., Berlin [94]
Thiel & Schuchardt, Metallwarenfabrik Gesellschaft mit beschränkter Haftung, Ruhla [L3]
Bernhard Thormann, Berlin [90]
Trolitan-Presswerk, [Ernst Meyer]Weiskirchen (Bez. Trier) [Z3]

[-U-]

Hein. Ulbricht's Witwe., Ges.m.b.H., Kaufing bei Schwanenstadt, Oberdonau [3X]

Karl Unger & Sohn, Metall- und Kunstharz-Presswerk, Gablonz a. Neiße (Reichsgau Sudetenland) [1Z]

[-V-]

Gebr. Vedder K.G., Schalksmühle i. Westf. [84]
[Vereinigte Chemische Fabriken Kreidl, Heller & Co., Nfg., Wien [VCF]]
Vereinigte Isolatorenwerke Aktiengesellschaft, (Viacowerke) Berlin-Pankow [31]
Vereingte Telefon- & Telegrafen-Werke Aktiengesellschaft, Wien 20 [0U]
Max Volkenrath, Ing., Wipperfürth (Rhld.) [X2]
Volkenrath und Co, Schwenke i. Westf. [A3]
Gebr. Vollmerhaus, Kierspe-Bahnhof i. Westf. [50]
Vossloh-Werke G.m.b.H., Werdohl i. Westf. [E2]

[-W-]

Wacker und Doerr, Nieder-Ramstadt bei Darmstadt [62]
Heinrich Wander, Gablonz a. Neiße (Reichsgau Sudetenland) [2L]
Karl Wegner, Berlin [T3]
Weisse und Co. Gräfental i. Thür. [V0]
Dr. Wellborn & Wernicke, Berlin [0A]
"Wellit" Gesellschaft Pless & Co., Kunstharz-Presswerk und Fabrik elektrotechnischer Installationsartikel, Wien [3V]
L. Adolf Werneburg, Sürth bei Köln [1L]
Westdeutsche Metallindustrie Wilhelm Kötter, Unna i. Westf. [T0]
Wester, Elbinghaus & Co., Hanau a. Main [F9]
Westfälische Metallwaren-Fabrik Christophery G.m.b.H., Iserlohn [2V]
Bruno Wetzstein, Plauen i. Vogtl. [Z2]
Casp[ar]. Arn[old]. Winkhaus, Carthausen i. Westf. [H1]
Erich Wippermann, Halver i. Westf. [46]
Wirth & Schirp, Presswerk, Rodt-Müllenbach über Marienheide (Bez. Köln) [3W]
Wolff & Co., Abt. Pressmaterial, Walsrode [33]

[-Z-]

Alois Zettler , Elektrot. Fabrik G.m.b.H, München [U3]
F.G. Zieger, Rosswein i. Sa. [0W]

Appendix V: Manufacturers by Two-Digit Number Designation

Names, firms, and places below were transcribed exactly as they were originally written in the wartime publications, particularly the 1940 *Bekanntmachung über Kunstharz-Preßmassen für typisierte und überwachte Preßstoffe* (Announcement Concerning Synthetic Resin Molding Compounds for Standardized and Inspected Moldings) as published in *Kunstoffe*, Issue 30.

In numeric order for reference, as cataloged and summarized by the author:

21 New-York Hamburger Gummi-Waaren Compagnie, Hamburg

22 Bebrit-Preßstofferke G.m.b.H., Bebra u. C. & F, Schlothauer G.m.b.H., Ruhla (Thür.); Bebra (H.-N.)

23 Busch-Jaeger Lüdenscheider Metallwerke Aktiengesellschaft, Lüdenschied i. Westf.

24 Gebrüder Merten, Gummersbach (Rhld.)

25 Süddeutsche Isolatoren-Werke G.m.b.H, Freiburg im Breisgau

26-30 (unexplained gap in sequence)

31 Vereinigte Isolatorenwerke Aktiengesellschaft, (Viacowerke) Berlin-Pankow

32 H. Römmler Aktiengesellschaft, Spremberg (Nd.-Lausitz)

33 Wolff & Co., Abt. Pressmaterial, Walsrode

34 Siemens-Schuckertwerke, Aktiengesellschaft, Abteilung Isolierstoffe (SK 4), Berlin-Siemensstadt (Gartenfeld)

35 Heliowatt-Werke, Elektrizitäts-Aktiengesellschaft, Berlin-Charlottemburg

36 Gebrüder Adt Aktiengesellschaft, Wächtersbach H.-N.

37 (not used)

38 Allgemeine Elektricitäts-Gesellschaft, Fabriken Hennigsdorf, Henningsdorf (Osthavelland)

Appendix V: Manufacturers by Two-Digit Number Designation

38A	Allgemeine Elektricitäts-Gesellschaft, Fabriken Annaberg, Annaberg i. Erzgeb. [Presswerk Scheibenberg i. Erzgeb] [62]
39	Dr. Deisting & Co., Gesellschaft mit beschrankter Haftung, Kierspe i. Westf.
40	Isola Werke A.G., Birkesdorf -Düren (Rhld.)
41-42	(unexplained gap in sequence)
43	Dynamit-Actien-Gesellschaft vormals Alfred Nobel & Co., Abteilung Celluloid-und-Kunstoff-Fabrik, Werk Troisdorf (Bez. Köln); verkauf durch Venditor, Kunststoff-Verkaufsges. m.b.H, Troisdorf (Bez. Köln)
44	(not used)
45	Presswerk A.G., Essen
46	Erich Wippermann, Halver i. Westf.
47-49	(unexplained gap in sequence)
50	Gebr. Vollmerhaus, Kierspe-Bahnhof i. Westf.
51	Presswerk Winkel, Schulte und Conze, Herscheid i. Westf.
52	(not used)
53	Ernst Backhaus & Co., Kierspe-Bahnhof i. Westf.
54	Ellinger & Geißler. Dorfhain (Bez. Dresden)
55	Robert Bosch G.m.b.H., Metallwerk, Stuttgart-Feuerbach
56	Radio H. Mende & Co., Abt. Presswerk "Mendelith", Dresden
57	(not used)
58	Deutsche Philips-Gesellschaft m.b.H., Berlin
59	Seckelmann & Co., Lüdenscheid i. Westf.
60	Oskar Gaudlitz, Coburg
61	Ernst Bremicker, Ing. Kierspe-Bahnhof i. Westf.

62	Wacker und Doerr, Nieder-Ramstadt bei Darmstadt
63-64	(unexplained gap in sequence)
65	Presswerk Königstein, R. Saring, Königstein, Sächs. Schweiz
66	Hermann Ros, Coburg
67	Bayerische Elektrozubehör G.m.b.H., Lauf bei Nürnberg
68	Mix & Genest Aktiengesellschaft, Berlin-Schöneberg
69	(not used)
70	Bisterfeld & Stolting, Inhaber: Ernst Bisterfeld, Radevormwald (Rhld.)
71	Wilhelm Geiger G.m.b.H, Lüdenscheid i. Westf.
72	Leopold Kostal, Lüdenscheid i. Westf.
73	Wilhelm Quante, Spezialfabrik für Apparate der Fernmeldetechnik, Inhaber: Hermann Quante, Wuppertal-Elberfeld
74	Pressmaterial-Werk Hermann Römmler und Schumann K.G., Berlin-Friedenau
75	(not used)
76	Bezet-Werk Hermann Buchholz, Motzen (Krs. Teltow)
77	(not used)
78	Hugo Krieger und Faudt, Berlin
79	Linden und Co. K.G., Lüdenscheid i. Westf.
80	Gebr. Berker, Schalksmühle i. Westf.
81	Pressstoffwerk Schöppenstedt, Paul Schnake, Schöppenstedt
82	Paul Teich, Berlin
83	Christian Geyer, Nürnberg
84	Gebr. Vedder K.G., Schalksmühle i. Westf.

Appendix V: Manufacturers by Two-Digit Number Designation

85	Theodor Krägeloh & Comp., Dahlerbrück i. Westf.
86	Kurt Steidel, Berlin
87	Paul Hochköpper und Co., Lüdenscheid I Westf.
88-89	(unexplained gap in sequence)
90	Bernhard Thormann, Berlin
91	(not used)
92	Futurit-Werk Aktiengesellschaft, Wien XI
93	Porzellanfabrik Bernhardshütte G.m.b.H., Blechhammer bei Sonneberg i. Thür.
94	Thega-Kontakt G.m.b.H., Berlin
95	Erich Jäger K.-G., Bad Homburg v.d. Höhe
96-97	(unexplained gap in sequence)
98	Elektrotechnische Fabrik Weber & Co., Komm.-Ges., Kranichfeld i. Thür.
99	Friedrich Dörscheln, Lüdenscheid i. Westf.

Appendix VI:
Manufacturers by Letter-Number Designation

Manufacturers by Letter-Number designations, as tabulated, cataloged, and summarized by the author:

A3	Volkenrath und Co, Schwenke i. Westf.
A4	Deutsche Legrit- Ges. m.b.H. Berlin
A8	Lindner und Co., Jecha-Sondershausen i.Thür.
B, C, D	prefixes apparently not used or not assigned to *Kunststoff* manufacturers
E0	Elektrotechnische Fabrik J. Carl, Gesellsch. m. beschr. Haftung, Oberweimar i. Thür.
E2	Vossloh-Werke G.m.b.H., Werdohl i. Westf.
E4	Paul Jatow, Dodendorf (bez. Magdeburg)
E7	Gebr. Spindler Betr.-Kom.-Ges., KG, Köppelsdorf i. Thür.
E8	Graewe und Co., Menden i. Westf. (Kreis Iserlohn)
E9	Lohmann und Welschehold, Meinerzhagen i. Westf.
F0	Otto Backhaus, Bollwerk i. Westf.
F1	Alusil-Preßstoffwerk Eugen Gassmann, Probstzella i. Thür.
F2	Franz Stauch, Presswerk, Unterrodach i. Ofr.
F4	Robert Anke, elektrot. Porzellanfabrik & Kunstharzpresserei, Oelsnitz i. Vogtl.
F5	Max Schulze, Meißen i. Sa.
F7	(does not appear on the 1940 list, but products bearing F7 are reported to exist)
F8	Gebr. Broghammer, Schramberg (Schwarzwald)

Appendix VI: Manufacturers by Letter-Number Designation

F9	Wester, Elbinghaus & Co., Hanau a. Main
G	prefix not used or not assigned to *Kunststoff* manufacturers
H1	Casp[ar]. Arn[old]. Winkhaus, Carthausen i. Westf.
H2	Emil Adolff, Abt. Kunstharz-Presswerk, Reutlingen i. Württ.
H3	Otto Langmann, Kunstharz-Presswerk, Hagen i. Westf.
H4	Gerdes & Co., Schwelm i. Westf.
H5	Hans Büllmann Werke für Elektrotechnik und Feinmechanik, Gablonz-Schlag (Reichsgau Sudetenland)
H7	Gebrüder von der Horst, Lüdenscheid i. Westf.
H8	Friemann & Wolf G.m.b.H., Zwickau i. Sa.
H9	Fabrik isolierter Drähte u. Schnüre, Schulze, Schneider und Dort G.m.b.H., Schönow, Post Bernau bei Berlin
I, J, K	prefixes apparently not used or not assigned to *Kunststoff* manufacturers
L0	Gebrüder Reiher K.G., vormals Aktiengesellschaft für Elektrotechnik, Braunschweig
L1	Hoppmann und Mulsow, Hamburg
L2	Isopress-Werk G.m.b.H., Berlin-Oberschöneweide
L3	Thiel & Schuchardt, Metallwarenfabrik Gesellschaft mit beschränkter Haftung, Ruhla i. Thür.
L4	Meirowsky & Co. Aktiengesellschaft, Porz a. Rhein.
L5	Porzellanfabrik Theodor Pohl, Schatzlar i. Riesengebirge
L6	Krone und Co, Berlin-Baumschulenweg
L7	Brökelmann, Jaeger & Busse, Neheim i. Westf.
L8	H. Bodenmüller, Ing., Stuttgart-Zuffenhausen
L9	Robert Karst, Berlin

M0	Carl Frier. Lübold, Lüdenscheid i. Westf.
M1	Bender & Wirth, Kierspe-Bahnhoff i. Westf.
M2	Siemens-Schuckertwerke, Aktiengesellschaft, Wiener Maschinen- und Apparate-Werk, Wien 20
M3	Kunstharz-Presserei Carl Germer, Berlin
M4	Kronacher Porzellanfabrik Stockhardt und Schmidt-Eckert, Kronach i. Bayern
M5	Richard Giersiepen, Bergisch Born (Rhein-Wupperkreis)
M6	Kaiser und Spelsberg, Schalksmühle i. Westf.
M7	Frankl & Kirschner, Elektrizitätsgesellschaft, Mannheim-Neckarau
M8	Elektrotechnische Metallwarenfabrik Storch und Stehmann G.m.b.H., Ruhla i. Thür.
M9	Jos. Mellert, Bretten i. Baden
N1	Sursum Elektr.-Gesellschaft Leyhausen & Co., Nürnberg
N2	Markische Elektro-IndustrieAdolf Vedder K.G., Schalksmühle i. Westf.
N4	Paul Jordan, Elekrotechnische Fabrik, Berlin-Steglitz
N5	Kunstharzpresserei M. Hildebrandt und E. Hammerschmidt, Brand-Erbisdorf i. Sa.
N6	Kurt Postel, Köln, Höhenberg
N7	Reicolit-Presswerk, Cuno Heinzelmann-Hassberg, Berlin
N8	Richard Brünner, Elektrotechnische Fabrik, Wien
N9	Karl Potthof, Presswerk, Solingen-Ohlings
0, P, Q, R, S, T	prefixes apparently not used or not assigned to *Kunststoff* manufacturers
T0	Westdeutsche Metallindustrie Wilhelm Kötter, Unna i. Westf.

Appendix VI: Manufacturers by Letter-Number Designation 145

T1	Gustav Branscheid & Co., Lüdenscheid
T2	Wilh. Burgbacher K.G., Neukirch i. Baden (Station Furtwangen)
T3	Karl Wegner, Berlin
T4	Bamburger Indistrie-Gesellschaft, Bamburg, i. Bayern
T6	Elektro-, Glimmer-und Preßwerke Scherb & Schwer K.-G. vorm. Jareslaw, Berlin-Weißensee
T7	Fritz Heublein, Neustadt bei Koburg
T8	Agalitwerk Milspe, Kattwinkel & Co., Milspe i. Westf.
T9	WG Petersen & Co., Wandsbeck
U1	"Feba" Fabr. Elektr. Bedarfsartikel Stückrath K.-G., Berlin-Köpenick
U2	Preh, Elektrofeinmechanische Werke, Bad Neustadt/Saale
U3	Alois Zettler , Elektrot. Fabrik G.m.b.H, München
U4	Apparatebauanstalt Schneider & Co. Breslau-Gr. Ohlewiesen
U5	Karl Buchruker, München
U6	Metallwerk Elektra G.m.b.H., Gummersbach (Rhld.)
U7	Wilhelm Sihn jr., Niefern i. Baden
U9	Ludwig Schröder, Schalksmühle i. Westf.
V0	Weisse und Co. Gräfental i. Thür.
V1	Phenoplast, Bischoff & Co., Kom.-Ges., Eberswalde
V2	Otto Nettelbeck, Berlin
V3	Goltzsche, Merlin & Sohn, Großröhrsdorf i. Sa.
V4	Bachmann & Leichsenring, Berlin-Neuköln
V5	Fr. Möller, Brackwede i. Westf.

V6	Fresen & Co., Lüdensheid i. Westf.
V7	Carl Walther, Waffenfabrik, Zella-Mehlis i. Thür.
V8	Signalapparatefabrik Julius Kräcker Aktiengesellschaft, Berlin
V9	Presstoffwerk Nürnberg, Gebrüder Klein, Nürnberg
W0	Kunstharz-Presserei Schwaben, Ingenieur Otto Single, Plochingen i. Württ.
W1	Heinrich Kopp G.m.b.H., Sonneberg i. Thür.
W2	Bonner Keramik Aktiengesellschaft, Bonn a. Rhein
W3	Dralowid-Werk der Steatit-Magnesia-Aktiengesellschaft, Teltow bei Berlin
W5	Strauss & Co., Schmölln i. Thür., Abteilung Kunstharzpresserei, Auma i. Thür.
W6	Blumberg & Co., Lintorf (Bez. Düsseldorf)
W7	Ernst Albert Senf, Kunstharzpresserei, Bautzen i. Sa.
W8	Heinrich Knöll, Gross-Bieberau i. Odenwald
W9	August Hessmert, Brügge
X0	Richard Rinker G.m.b.H., Menden (Krs. Iserlohn)
X1	W. Stiefelding, Berlin, SO 36
X2	Max Volkenrath, Ing., Wipperfürth (Rhld.)
X3	Presswerk Kurt Wentzel, Inhaber Anton Jacob, Berlin-Steglitz
X4	G. Rövenstrunk & Co., Elspe bei Brügge i. Westf.
X5	Ostland G.m.b.H., Königsberg i. Pr.
X6	Julius Karl Görler, Transformatorenfabrik, Berlin-Charlottenburg
X7	Kugella vormals Max Roth G.m.b.H., Mittelschmalkalden (Post Wernshausen)
X8	Ernst Gomolka, Zehdenick (Mark)

Appendix VI: Manufacturers by Letter-Number Designation 147

X9	Ernst Gösser, Iserlohn i. Wesf.
Y0	Herbert Horn, Pulsnitz i. Sa.
Y1	Ferdinand Schuchhardt, Berliner Fernsprech-und Telegraphewerk A.G., Berlin
Y3	Goseberg & Grashoff, Kierpe-Bahnhof i. Wesf.
Y5	Schmachtenberg & Türck, Solingen-Wald
Y7	Fischer & Klüppelberg vorm. W. Dörner & Co., G.m.b.H., Radevormwald (Rhld.)
Y8	Adolf Ruoff, Kunstoff-Presswerk, Radevormwald (Rhld.)
Y9	C. Lorenz Aktiengesellschaft, Berlin-Tempelhof
Z0	Plate & Voerster, Kierspe i. Westf.
Z1	Internationale Galalith-Gesellschaft Hoff & Co., Hamburg-Harburg
Z2	Bruno Wetzstein, Plauen i. Vogtl.
Z3	Trolitan-Presswerk, Weiskirchen (Bez. Trier)
Z4	Paul Kuhbier & Co., Wipperfürth (Rhld.)
Z5	Eisele Auto-Electric, Inhaber Karl Klos, Frankfurt a. Main
Z6	Gebr. Dahlhaus, Schalksmühle i. Westf.
Z7	Heinrich Schmidberger, Wien
Z8	Rheinisch-Westfälisches Kunststoffwerk G.m.b.H., Mülheim-Ruhr
Z9	Köditz & Co., Metallwarenfabrik u. Presserei, Langeweisen i. Thür.

Appendix VII:
Manufacturers by Number-Letter Designation

Manufacturers by Number-Letter designations, as tabulated, cataloged, and summarized by the author:

0A	Dr. Wellborn & Wernicke, Berlin
0E	Metallwerke Adolf Hopf, Aktiengesellschaft, Tambach-Dietharz i. Thür.
0F	C. Pose, Wehrausrüstungen, Berlin
0H	Erlemann & Co., Bergerhof (Rhld.)
0L	Kunstharz-Presswerk der Hanf-, Jute- u. Textil-Industrie A.G., Wien
0M	Dornseif & Linde, Kierspe i. Westf.
0N	Dr. Paul Isphording, Kunstharzpresserei, Wepritz bei Landsberg (Warthe)
0T	Barth, Klemm & Co., Leipzig
0U	Vereingte Telefon- & Telegrafen-Werke Aktiengesellschaft, Wien
0V	Presswerk Mollberg & Co., Hofgeismar, Bez. Kassel
0W	F.G. Zieger, Rosswein i. Sa.
0X	Karl Friedr. Seiter K.-G., Bollwerk i. Westf., Post Oberbrügge
0Y	Ing. Fr. August Pfeffer, Oberlind i. Thür.
0Z	Acla, Rheinische Maschinen-, Leder- u. Riemenfabrik, Aktiengesellschaft, Köln Mülheim
1A	Karl Müller, Solingen-Merschied
1E	Schieck-Instrument Wilhelm Wolkersdorf, Berlin
1F	Singer Nähmaschinen Aktiengesellschaft, Fabrik Wittenberge, Bez. Potsdam
1H	F.F.A. Schulze, Metallwarenfabrik, Berlin

Appendix VII: Manufacturers by Number-Letter Designation

1L	L. Adolf Werneburg, Sürth bei Köln
1M	Gebrüder Kaiser & Co., Neheim-Ruhr
1N	Böhmische Kontactwerke Aktiengesellschaft, Komotau (Reichsgau Sudetenland)
1T	Hans Heußinger, Ing., Kunstoff-Presserei, Nürnberg-N
1U	Johannes Buchsteiner, Gingen-Fils i. Württemberg
1V	Ostpreussisches Kunstharz-Presswerk, Gedenk, Blank & Naujoks-Erben, Königsberg i. Pr.
1W	Julius Posselt, Gablonz a. Neiße (Reichsgau Sudetenland)
1X	Hering & Co., Iserlohnerheide bei Iserlohn, Post über Schwerte/Ruhr
1Y	Grelit Kunstharz-Presswerk Grohmann, Pietschmann & Co., Nixdorf (Reichsgau Sudetenland) [cws]
1Z	Karl Unger & Sohn, Metall- und Kunstharz-Presswerk, Gablonz a. Neiße (Reichsgau Sudetenland)
2A	Carl Pfestorf, Abt. Kunstharzpresserei, Tambach-Dietharz i. Thür.
2E	Josef Feix Söhne, Gablonz a. Neiß (Reichsgau Sudetenland)
2F	Franz Kirsten, Elektrotechische Spezialfabrik, Bingerbrück a. Rhein
2H	Brunnquell & Co., Sonderhausen i. Thür.
2L	Heinrich Wander, Gablonz a. Neiße (Reichsgau Sudetenland)
2M	Richard Hirschmann, Esslingen a. N.
2N	Robert Friedrich, Abteilung III [Kartonagenfabrik], Annaberg i. Erzgeb.
2T	Starkstrom-Apparatebau G.m.b.H., Berlin; Presserei: Zweigwerk Buschullersdorf bei Reichenberg (Reichsgau Sudentenland)
2U	Gerhard Lischke, Kunstharzpresserei, Oppach i. Sa.
2V	Westfälische Metallwaren-Fabrik Christophery G.m.b.H., Iserlohn

2W	Schriever & Co., Kunstoffpresserei, Kierspe-bahnhof i. Westf.
2X	Bergfeld & Heider, Burscheid (Bez. Düsseldorf)
2Y	Ernst Klever, Kunstharz-Presserei, Solingen-Nord
2Z	Röhrenwerk Johannes Surmann G.m.b.H., Arnsberg i. West.
3A	Max Braun, Frankfurt a. Main
3E	Schmidt & Co. K.G., Metallwarenfabrik, Schwelm i. Westf.
3F	Reininghaus & Co., Beleuchtungskörper-Fabrik, Lüdenscheid i. Westf.
3H	Willi Kleinau, Kunstharz-Preß- und Spritzwerk, Berlin
3L	Gebr. Sieling, Metallwaren-Fabrik und Kunstharz-Presswerk, Lüdenscheid i. Westf.
3M	Anton Reiche A.-G., Dresden
3N	Telefonbau und Normalzeit G.m.b.H., Franfurt a. Main
3T	H. Merlet, Kunststoffpresswerk, Neudorf b. Glabonz a. Neiße (Reichsgau Sudetenland)
3U	George Rüger & Co., vorm. Elektra G.m.b.H., Elektrotechnische Fabrik, Essen-Heidhausen
3V	"Wellit" Gesellschaft Pless & Co., Kunstharz-Presswerk und Fabrik elektrotechnischer Installationsartikel, Wien
3W	Wirth & Schirp, Presswerk, Rodt-Müllenbach über Marienheide (Bez. Köln)
3X	Hein[rich]. Ulbricht's Witwe., Ges.m.b.H., Kaufung bei Schwanenstadt, Oberdonau
3Y	Ludwig Richter, elektrotechn. Spezialfabrik, Görlitz

Appendix VII: Manufacturers by Number-Letter Designation

Of note, the author has observed number-letter combinations beyond those listed in the 1940 listing as published in *Kunstoffe*. This lends credence to the assumption that subsequent listings probably exist and that companies that commenced operations after 1940 were likely assigned these currently unconfirmed designators. A list of these as observed by the author follow:

[4M Seen on black canteen cup made from "S" material]

[4Z "List" brand, seen on Luftwaffe electrical components]

[6O Late black butter dish marked "S"]

[6F Late black phenolic canteen cup marked "6F" and "S" on 44 "cfl" canteen]

Appendix VIII:
Military Letter Codes Assigned to *Kunststoffe* Manufacturers

The majority of firms listed and supervised by the MPD were probably eventually assigned a one, two, or three digit military code by the OKH to replace their MPD numeric or alphanumeric designations. It would not be practical to re-list every firm already documented in the previous appendices with their wartime military letter codes, as numerous publications covering these codes for the whole of Germany are still in print that the reader can reference. The author has though organized and where possible, cross-referenced in brackets, the most prominent and most often observed markings, which follows below for quick reference.[63]

Of note, a number of firms in occupied countries that made *Kunststoffe* under German supervision are also listed by code, do not appear to have been assigned an MPD designation by the *Staatliches Materialprüfungsamt* in Berlin-Dahlem.

ac	Carl Walther Waffenfabrik, Zella-Mehlis, Thüringen [V7]
acb	Rheinisch-Westfälisches Kunstoffwerk G.m.b.H., Mülheim-Ruhr [Z8]
adh	C.A. Winkhaus Elektrotechnische Bedarfsartikelfabrik [H1]
af	Günther Wagner, Hannover
ajg	Erich Jäger K.-G., Bad Homburg v.d. Höhe [95]
ajs	Christian Geyer, Nürnberg [83]
amh	Hans Büllmann Werke für Elektrotechnik und Feinmechanik, Gablonz-Schlag (Reichsgau Sudetenland) [H5]
ans	Hein[rich]. Ulbricht's Witwe., Ges.m.b.H., Kaufung bei Schwanenstadt, Oberdonau [3X]
ao	Heinrich Kopp G.m.b.H., Sonneberg i. Thür. [W1]
apr	Bisterfeld & Stolting, Isolierpreßstoffe, Werk Heinersdorf/Tafelfichte [subsidiary of 70]
aq	Plate & Voerster, Eisenwarenfabrik, Kierspe i. Westf. [Z0]
aqr	Singer Nähmaschinen Aktiengesellschaft, Fabrik Wittenberge, Bez. Potsdam [1F]

Appendix VIII: Military Letter Codes

awl	Union-Gesellschaft für Metallindustrie, Sils van de Loo & Co., Fröndberg/Ruhr
ay	Alois Pirkl, Elektrotechnische Fabrik, Reichenberg (Liberec)
ayf	Erfurter Maschinenfabrik Berthold Geipel G.m.b.H. (ERMA)
azg	Siemens-Schuckert-Werke, Aktiengesellschaft, Abteilung Isolierstoffe (SK 4), Berlin-Siemensstadt (Gartenfeld) [34]
azh	Siemens-Schuckert-Werke, Aktiengesellschaft, Abteilung Isolierstoffe (SK 4), Berlin-Siemensstadt (Gartenfeld) [34]
bbc	Mix und Genest Aktiengesellschaft, Berlin-Schöneberg [68]
bdl	"Isola"-Werke A.G., Birkesdorf -Düren (Rhld.) [40]
beh	Ernst Leitz, G.m.b.H., Optische Werke, Wetzlar
bfn	New York-Hamburger Gummi-Waaren Compagnie, Abt. Kunststoffe, Hamburg [21]
bgw	Alois Zettler , Elektrot. Fabrik G.m.b.H, München [U3]
bjp	Robert Bosch G.m.b.H., Metallwerk, Stuttgart-Feuerbach [55]
bko	"Micki", Dr. E. & H. Mickenhagen, Kunstharzpreßteile, Radevormwald
bkp	Röhrenwerk Johannes Surmann G.m.b.H., Arnsberg i. West. [2Z]
bl	Radio H. Mende & Co., Abt. Presswerk "Mendelith", Dresden [56]
blb	Lindner und Co., Jecha-Sondershausen i.Thür. [A8]
blc	Carl Zeiss, Optische Geräte / Militärabteilung, Jena
bmt	Optische Werke C. A. Steinheil Söhne G.m.b.H., München
boa	Venditor Kunststoff-Verkaufges m.b.H., Troisdorf [43]
bou	Telefunken Gesellschaft für drahtlose Telegraphie m.b.H., Werke-Erfurt & Berlin-Zehlendorf
bq	Brandt, Roland, Fabrik für Radiotelefonie, Berlin
bqa	Vereingte Telefon- & Telegrafen-Werke Aktiengesellschaft, Wien [0U]

brb	Richard Rinker, GmbH, Menden (Krs. Iserlohn) [X0]
brw	Schmake & Kumpmann A.G., Isolier- u. Stahlpanzerrohre, Hagen
btn	Ernst Gösser, Metallwarenfabrik u. Kunstharzpreßwerk, Iserlohn i. Westf. [X9]
bts	Matthias Oechsler & Sohn A.G., Riegersdorf/Sudetengau
btt	Robert Kreisel, Gürtler, Gablonz/Neiße
bug	Süddeutsche Telefon-Apparate-, Kabel-und Drahtwerke A.G., Nürnberg
bvu	Franz Kuhlmann, Werkstätten für Präzisionsmechanik u. Maschinenbau, Zentrale Wilhelmshaven-Rüstringen
bzh	Josef Mellert, Bretten/Baden [M9]
bzz	AGFA; I.G. Farben-Industrie AG, Werk "Agfa", München
cdu	Emil Adolff, Abt. Kunstharz-Presswerk, Reutlingen i. Württ. [H2]
clk	F.W. Breithaupt & Sohn, Kassel
cjy	"Neag" Norddeutsche Elektro-Akustik-Gesellschaft, Kaufhold KG, Berlin
cme	Gebr. Wichmann, Zeichengeräte, Vermessungsinstrumente, Berlin
cnn	Robert Karst, Berlin [L9]
cny	C. Pose, Wehrausrüstungen, Berlin [0F]
cpk	Internationale Galalith-Gesellschaft Hoff & Co., Hamburg-Harburg [Z1]
cqp	Signalapparatefabrik Julius Kräcker Aktiengesellschaft, Berlin [V8]
cqw	Bruno Wetzstein, Kunstharzartikel u. Presserei, Plauen i. Vogtl. [Z2]
cqx	Friemann & Wolf G.m.b.H., Zwickau i. Sa. [H8]
ctd	Anton Reiche A.-G., Dresden [3M]
cuy	Emil Adolff, Abt. Kunstharz-Presswerk, Reutlingen i. Württ. [H2]
cwh	Krone und Co, Berlin-Baumschulenweg [L6]

Appendix VIII: Military Letter Codes

cws	"Grelit" Kunstharz-Presswerk Grohmann, Pietschmann & Co., Nixdorf (Reichsgau Sudetenland) [1Y]
cwu	Georg Kremp Aktiengesellschaft für Optik u. Mechanik, Wetzlar
cxn	Emil Busch A.G., Rathenow
czp	Ostland G.m.b.H., Königsberg i. Pr. [X5]
dbf	Hoppmann & Muslow, Hamburg [L1]
ddf	Lohmann-Werke A.G., Bielefeld
ddu	Richard Bosse & Co., Telegrafenbauanstalt, Berlin
ddx	Voigtländer & Sohn AG, Optisches Gerät, Werk Brauschweig
ded	Heliowattwerke, Elektrizitäts-Aktiengesellschaft, Berlin-Charlottemburg [35]
dfy	Porzellanfabrik Bernhardshütte G.m.b.H., Blechhammer, bei Sonneberg i.Thür. [93]
dju	Wedig & Reuss, Kunstharz-Fabrik, Eilenburg/Mulde
dkn	Wilhelm Geiger G.m.b.H, Lüdenscheid i. Westf. [71]
dlk	Gebrüder Adt, Aktingesellschaft, Wächtersbach [36]
dmr	C. Lorenz Aktiengesellschaft, Berlin-Tempelhoff [Y9]
dnb	Borck & Goldschmidt, Werkstatt f. Telegrafie u. Telefonie, Berlin
dnf	Rheinische-Westfalische Sprengstoff Actiengesellschaft, Werk Stadeln, bei Fürth [43]
dnz	"Saba" Schwarzwälder Apparate-Bauanstalt, August Schwer Söhne, Villingen
dom	"Hella" Westfälische Metall Industrie KG Hueck & Co., Lippstadt
dpa	Elektro-Isolier-Industrie Wahn, Wilhelm Ruppert, Köln-Porz-Wahn
dsk	Reppel & Vollmann, Kierspe
dtf	Preh, Elektrofeinmechanische Werke, Bad Neustadt/Saale [U2]

dum	Otto Sturm, Kunstharz-Preßwerk, Werkzeugbau, Wuppertal-Barmen
dxa	Ellinger & Geißler. Dorfhain (Bez. Dresden) [54]
eaa	Telefonbau und Normalzeit G.m.b.H., Franfurt a. Main [3N]
eas	Siemens & Halske AG, Berlin-Siemensstadt
ecq	Ferdinand Schuchhardt, Berliner Fernsprech-und Telegraphewerk A.G., Berlin [Y1]
edv	Thiel & Schuchardt, Metallwarenfabrik Gesellschaft mit beschränkter Haftung, Ruhla i. Thür. [L3]
eep	Brökelmann, Jaeger & Busse, Neheim i. Westf. [L7]
efp	Ernst Backhaus & Co., Kunstharzpresserei, Kierspe-Bahnhof i. Westf. [53]
efq	Bachmann & Leichsenring, Isoliermaterial, Berlin-Neukölln [V4]
efv	Bisterfeld & Stolting, Inhaber: Ernst Bisterfeld, Radevormwald (Rhld.) [70]
efy	Elektrotechnische Metallwarenfabrik Storch & Stehmann G.m.b.H., Ruhla [M8]
egf	Fritz Heublein [Witwe], Neustadt bei Coburg [T7]
egq	Kunstharz-Presserei Schwaben, Ingenieur Otto Single, Plochingen i. Württ. [W0]
egx	"Phenoplast" Bischoff & Co. KG, Kunstharzartikel u. Presserei, Eberswalde-Finow [V1]
egy	Ing. Fr. August Pfeffer, Sonneberg-Oberlind [0Y]
egz	Kurt Postel, Köln, Höhenberg [N6]
eha	"Reicolit"-Presswerk Cuno Heinzelmann-Hasberg, Kunstharzpresserei, Berlin [N7]
ehe	H. Römmler Aktiengesellschaft, Spremberg Nd.-Lausitz [HRS/32]
ehj	Gebrüder Spindler Betr.-K.G., Köppeldorf i. Thür. [E7]
eja	Richard Schneider, Werk Coburg

Appendix VIII: Military Letter Codes

eje	Blumberg & Co. Bürobedarfsartikelfabrik, Lintorf [W6]
ejo	Presswerk A.-G. Kunstharzartikel, Essen [45]
eky	Volkswagenwerk GmbH, Stadt des KdF-Wagens bei Fallersleben
eme	Dr. Deisting & Co., Gesellschaft mit beschrankter Haftung, Kierspe i. Westf.[39]
emq	Willi Stiefeling, Kunstharzpresserei, Berlin [X1]
enl	Allgemeine Elektricitäts-Gesellschaft, Fabriken Hennigsdorf, Henningsdorf (Osthavelland) [38]
eor	Günther Wagner, Verpackungswerk, Hannover-Heinholz
eot	Fichter & Hackenjos K.G., Fabrik für Feinmechanik, Villingen/Schwarzwald
epl	Lytax-Werke G.m.b.H., Freiburg/Breisgau
epm	G. Schanzenbach & Co. G.m.b.H., Elektro- u. Lichttechnische Spezialfabrik, Frankfurt/Main
ept	Hamar Jernstöperi & Mek. Verksted, Hamar
epy	Presstoffwerk Schöppenstedt, Paul Schnake, Schöppenstedt [81]
epz	Elektrotechnische Fabrik Weber & Co., Komm.-Ges., Kranichfeld i. Thür. [98]
eqd	Futurit-Werk A.G., Wien [92]
eqe	Heinrich Schmidberger, Fabrikation von Artikeln aus Kunststoff, Wien [Z7]
eqo	Robert Friedrich, Abteilung III [Kartonagenfabrik], Annaberg i. Erzgeb. [2N]
eqq	KACO; Gustav Bach Kupfer-Asbest-Co., Heilbronn
eqt	Preßmaterial-Werk Hermann Römmler & Schumann Komm.-Ges., Berlin-Friedenau [74]
eqw	Carl Pfestorf, Abt. Kunstharzpresserei, Tambach-Dietharz i. Thür. [2A]
erz	Kronacher Porzellanfabrik Stockhardt & Schmidt-Eckert, Kronach i. Bayern [M4]

esq	NOTEK, Nova-Technik G.m.b.H, München
etl	Hugo Krieger & Faudt, Fabrik elektrischer Apparate, Berlin [78]
esu	Optische Werke C. A. Steinheil Söhne G.m.b.H., München
ewu	Gebrüder Kaiser & Co., Neheim-Ruhr [1M]
fef	Batterien- u. Elementen-Fabrik, System Zeiler AG, Berlin
fhw	Solmonit Fabrik für Isolierstoffe der Elektrotechnik, Sonneberg
flf	Vossloh-Werke G.m.b.H., Werdohl i. Westf. [E2]
fnb	Gebrüder Reiher K.G., vormals Aktiengesellschaft für Elektrotechnik, Braunschweig [L0]
fnk	Metallwerke Adolf Hopf, Aktiengesellshaft, Tambach-Dietharz i. Thür. [0E]
fnm	Wiener Isolierrohr-, Batterie- u. Metallwarenfabrik G.m.b.H., Wien
fnq	Rötelmann & Co. KG, Armaturenfabrik u. Fassondreherei, Werdohl
fnu	Bergfeld & Heider, Burscheid (Bez. Düsseldorf) [2X]
foc	"Bebrit"-Preßstoffwerke G.m.b.H., Bebra u. C. & F, Schlothauer G.m.b.H., Ruhla (Thür.); Bebra (H.-N.) [22]
fpc	Isopress-Werk G.m.b.H., Berlin-Oberschöneweide [L2]
fpb	Hartmann & Braun AG, Elektrische & wärmetechnische Meßgeräte, Frankfurt
fpr	Fr. Möller, Brackwede i. Westf. [V5]
fpr	Fr. Möller, Kunstharzpreßwerk, Brackwede i. Westf. [V5]
fqv	Gesellschaft f. elektr. Unternehmungen AG: Fabrik Loewe-Gesfürel AG, Berlin
fvw	Blaupunkt-Werke G.m.b.H., Berlin-Wilmersdorf
fwf	Josef Feller, Fabrik isolierter Drähte u. Kabel, Wien
fxn	Schieck-Instrumente, Wilhelm Wolkersdorf, Berlin [1E]

Appendix VIII: Military Letter Codes 159

fyb	Franz Stauch, Presswerk, Unterrodach i. Ofr. [F2]
fzd	Ernst Pless, Werkstätte für photografische. Artikel, Wien
fzn	Hugo Krieger und Faudt, Berlin [78]
fzz	Köditz & Co., Metallwarenfabrik u. Presserei, Langeweisen i. Thür. [Z9]
gbj	Erich Trost, Isolan-Preß-Fabrik, Berlin
gbk	Dr. Wellborn & Wernicke, Kunstharz-Presserei u. –Spritzerei, Berlin [0A]
gbm	Vereinigte Isolatorewerke A.G., (Viacowerke), Berlin-Pankow [31]
gdr	Gebrüder Merten, Gummersbach (Rhld.) [24]
gds	Gebr. Dahlhaus, Holzwarenfabrik, Kunstharzpreßwerk, Schalksmühle i. Westf. [Z6]
geu	Paul Kuhbier & Co., Fabrik für Präzisionspreßstücke, Wipperfürth [Z4]
GEU	Paul Kuhbier & Co., Fabrik für Präzisionspreßstücke, Wipperfürth [Z4]
gfb	Gerdes & Co., Spezialfabrik für Preßteile aus Kunststoffen, Schwelm i. Westf. [H4]
gfc	Julius Posselt, Gablonz a. Neiße (Reichsgau Sudetenland) [1W]
gfe	Elektro-, Glimmer und Preßwerke Scherb & Schwer K.-G. vorm. Jareslaw, Berlin-Weißensee [T6]
gfd	Porzellanfabrik Theodor Pohl, Schatzlar i. Riesengeb. [L5]
gfe	Elektro-, Glimmer und Preßwerke Scherb & Schwer K.-G. vorm. Jareslaw, Berlin-Weißensee [T6]
ggf	Meirowsky & Co. AG, Köln-Porz
ggh	Robert Schneider KG, Fabrik elektrotechnischer Apparate, Offenbach/Main
ggj	Gebr. Berker, Schalksmühle i. Westf. [80]
gkz	Leopold Kostal, Lüdenscheid i. Westf. [72]
gmv	Lynenwerk KG, Fabrik isolierter Drähte u. Kabel, Eschweiler

Kunststoffe

gol	Voss & Sohn, Fabrik für Kunstharzpreßteile, Lenhausen, bei Finnentrop/Lenne
gtr	F.F.A. Schulze, Metallwarenfabrik, Berlin [1H]
gun	Apparatebauanstalt Schneider & Co. Breslau-Gr. Ohlewiesen [U4]
guu	Schmachtenberg & Türck, Solingen-Wald [Y5]
gwq	Fabrik isolierter Drähte u. Schnüre, Schulze, Schneider & Dort G.m.b.H., Schönow, bei Bernau [H9]
gyh	Blei-Industrie A.G. (eh. Jung & Lindig), Freiberg
gyz	Dralowid-Werk der Steatit-Magnesia-Aktiengesellschaft, Teltow bei Berlin [W3]
hae	Goseberg & Grashoff, Kierpe-Bahnhof i. Wesf. [Y3]
hal	Wilhelm Quante, Spezialfabrik für Apparate der Fernmeldetechnik, Inhaber: Hermann Quante, Wuppertal-Elberfeld [73]
hap	Max Kohl AG, Phys. Apparate u. Laboratoriumseinrichtungen, Chemnitz
hbu	Elektro-Mechanik Heinrich List, Berlin-Teltow [4Z]
hdq	Carl Frier. Lübold, Lüdenscheid i. Westf. [M0]
hhb	Kontactwerke Aktiengesellschaft, Komotau (Reichsgau Sudetenland) [1N]
hkf	Josef Feix Söhne, Gablonz a. Neiß (Reichsgau Sudetenland) [2E]
hko	Matthias Oechsler & Sohn A.G., Metallwarenfabriken, Ansbach
hou	Ernst Bremicker, Ing. Kierspe-Bahnhof i. Westf.[61]
hpb	Presswerk Königstein, R. Saring, Königstein, Sächsische Schweiz [65]
hpx	Heliowatt-Werke, Elektrizitäts-Aktiengesellschaft, Berlin-Charlottenburg 4 [35]
hqn	Volkenrath und Co, Schwenke i. Westf. [A3]
htu	Strauss & Co., Schmölln i. Thür., Abteilung Kunstharzpresserei, Auma i. Thür. [W5]

Appendix VIII: Military Letter Codes

hyt	H. Römmler Aktiengesellschaft, Spremburg (Nd.-Lausiz) [HRS]
jba	A. Wunderlich Nachf., Heeres-, Polizei- u. Feuerwehrausrüstungen, Berlin-Neukölln
jda	Ing. Ferdinand Indruch, Apparatebau u. Kunstharzpreßwerk, Troppau/Sudetengau
jhy	Allgemeine Elektricitäts-Gesellschaft, Fabriken Henningsdorf, Henningsdorf (Osthavelland) [38]
jju	Paul Teich, Berlin [82]
jjw	Wirth & Schirp, Presswerk, Rodt-Müllenbach über Marienheide (Bez. Köln) [3W]
jlb	Stevens, St.Nicolas, Antwerpsche Steenweg
jlc	Usines Vynckier Freres & Co. SA, Gand (Ghent), Belgium
jld	Georges Leyers, Lüttich (Liége), Belgium
jle	Établissements Alfred Barth, Brüssel (Bruxelles), Belgium
jmb	Lederwaren-Industrie Stefanski, Posen
jmh	Heinrich Kopp, Inh. Theodor Simoneit, Sonnberg i/Thür. [W1]
jml	Karl Unger & Sohn, Metall- und Kunstharz-Presswerk, Gablonz a. Neiße (Reichsgau Sudetenland) [1Z]
jnc	Fritz Berges, Kunstharz-Erzeugnisse (in Griemeringhausen), Marienheide/Sauerland
jnd	Dornseif & Linde, Kunstharzpresserei, Kierspe i. Westf. [0M]
jne	Preßwerk Mollberg & Co., Hofgeismar/Esse
jnf	Süddeutsche Isolatoren-Werke G.m.b.H, Freiburg [25]
jng	"Trolitan"-Presswerk, Ernst Meyer, Weiskirchen, bei Losheim (Saar) [Z3]
jnm	Hartmann & Braun AG, Elektrische & wärmetechnische Meßgeräte, Werk Braunschweig

jnq	Joseph Müller, Kunstharzpreßwerk u. Werkzeugbau, Nürnberg
joe	Richard Brünner, Elektrotechnische Fabrik, Wien [N8]
jov	Presswerk Kurt Wentzel, Inhaber Anton Jacob, Berlin-Steglitz [X3]
jpa	Cogebi, Compagnie Generale Belge D'Isolants et Des Produits Moulos, Loth, bei Brüssel (Bruxelles) Belgium
jrx	Otto Kittel, Metallwaren, Kunstharzpresserei, Lusdorf/Tafelfichte (Ludvíkov pod Smrkem), Czechoslovakia
jva	E. Calsbach, Kunstharzpreßteile, Bergneustadt
jvg	Ernst Albert Senf, Kunstharz-Presserei, Bautzen i. Sa. [W7]
jwl	H. Merlet, Kunststoffpresswerk, Neudorf b. Gablonz a. Neiße (Reichsgau Sudetenland) [3T]
jwv	Karl Friedrich Selter K.-G., Bollwerk i. Westf., Post Oberbrügge, bei Halver [0X]
jww	Rohrbach & Co., Kunstharzpresserei, Lüdenscheid
jxm	Albert Nestler AG, Lahr/Schwarzwald
jyh	Kunstharz-Preßwerk u. Textilit Industrie A.G., Wien
jzp	Drucks & Co., Kunstharzartikel, Oberbrügge, bei Halver
kad	A.W. Faber-Castell Bleistiftfabrik AG, Werk Geroldsgrün, bei Schwarzenbach/Wald
kbs	Oskar Stärz, Fabrikation von Massenartikeln aus Kunstharz, Zittau
kcf	Richard Simm & Söhne, Metall- u. Kunststoffwarenfabrik, Prag, Czechoslovakia
kfe	Dietrich Lüttgens, Kunststoffpresserei, Velbert
kfx	Presswerk Winkel, Schulte und Conze, Herscheid i. Westf. [51]
kgo	Wilhelm Sihn jr., Niefern i. Baden [U7]
kjj	Askania-Werke, Berlin-Friedenau

kmb	Bellmann & Co., Kunstharzpreßwerk, Berlin
kqx	Karl Potthoff, Presswerk Solingen-Ohligs [N9]
lkd	Gerhard Lischke, Kunstharzpresserei, Oppach i. Sa. [2V]
mbv	AGFA; I.G. Farben-Industrie AG, Werk "Agfa", Berlin
mpm	Robert Bosch G.m.b.H., Metallwerk, Stuttgart-Feuerbach [55]
mqc	Siemens & Halske AG, Berlin-Siemensstadt
mqe	OSRAM G.m.b.H., Glühlampenfabrik, Berlin
mwx	Robert Bosch G.m.b.H., Metallwerk, Stuttgart [55]
mzj	Franz Schneider, Kunstharzpresserei, Kronach
ndm	Entwicklungsunternehmen Otto Seifert, Auma, bei Triptis
nhu	Luftfahrtgerätewerk Hakenfelde G.m.b.H., Berlin-Spandau
nht	Siemens-Schuckertwerke, Aktiengesellschaft, Abteilung Isolierstoffe (SK 4), Berlin-Siemensstadt (Gartenfeld) [34]
nnz	Philips; Deutsche-Philips Gesellschaft m.b.H., Berlin [58]
noa	Hermann Klasing KG, Draht- u. Isolierfabrik, Berlin-Adlershof
noy	Karl Buchrucker, Kunstharzpresserei, München [U5]
npc	"Dielektra" AG, Isoliermaterialien für Elektrotechnik, Köln-Porz
nqp	Goltzsche Merlin & Sohn Kunstharzpresserei, Großröhrsdorf i. Sa. [V3]
nqv	Ewald Hedfeld, Kunstharzartikel, Kierspe, bei Meinerzhagen/Sauerland
nvf	Ernst Steger & Co., Inh. Ernst Steger, Kunstharzartikel, Radevormwald
nvr	Gebr. Vollmerhaus, Kierspe-Bahnhof, i. Westf. [50]
nvv	Karl Wegner, Fabrik für Preßstoff-Massenartikel u. Isolierteile, Berlin [T3]
nwh	Richard Hirschmann, Esslingen a. N. [2M]

Kunststoffe

nwk	Elektro-Mechanik Heinrich List, Berlin-Teltow [4Z]
obj	Ernst Gösser, Metallwarenfabrik u. Kunstharzpreßwerk, Werk Niklasdorf (Mikulovice), [a subsidiary of X9]
ocw	Elektro-Mechanik Heinrich List, Berlin-Teltow [4Z]
ofa	Paul Naumann, Fabrik für Isolier-Preßteile, Berlin
off	Bohle & Cie. GmbH, Isolierwerk u. Metallwarenfabrik, Werk Köln-Mühlheim
ofh	Metall- u. Kunstharzwerk G.m.b.H., Komotau/Sudetengau (Chomutov)
ohj	Établissements Renau, Charlieu/Loire, bei Roanne, France
opn	Aug. Nowack Aktiengesellschaft, Bautzen [No]
otj	Kunstharz-Presserei Carl Germer & Willems, Berlin
qyh	unknown (late war code change, observed on *Kunststoff* MG34 butt)
rfd	unknown (late war code change, observed on *Kunststoff* rifle grenade components)
rln	Carl Zeiss, Optische Geräte / Militärabteilung, Jena
sbe	unknown (late war code change, observed on *Kunststoff* wrist compass)
t	"DAG" Dynamit AG (vorm. Alfred Nobel & Co.), Werk Troisdorf [43]
tpc	unknown (late war code change, observed on *Kunststoff* firearms components)
trf	Dynamit-Actien-Gesellschaft vormals Alfred Nobel & Co., Abteilung Celluloid-und-Kunststoff-Fabrik, Werk Troisdorf (Bez. Köln); verkauf durch Venditor, Kunststoff-Verkaufsges. m.b.H, Troisdorf (Bez. Köln) [t, DAG, 43]
wal	Wolff & Co., Abt. Pressmaterial, Walsrode [33]
xa	Busch-Jaeger Lüdenscheider Metallwerke Aktiengesellschaft, Lüdenscheid i. Westf. [23]

Glossary and Abbreviations

a.	Abbreviation for *"am" or "an"...*"; or "on the..."
Abnahme	Inspections Office
Abt.	Abbreviation for *Abteilung*; or detachment/subsidiary
Abteilung	Detachment, or subsidiary
A.G.	Abbreviation for *Aktiengesellschaft*; a "stock company", or roughly "incorporated"
Aktiengesellschaft	(as above)
am	"on..."
Anforderungszeichen,	"Standardized Requisition Number", roughly translated
BAL	Abbreviation for *Bauaufsicht Luft*, or Air force Construction Supervision
b.	Abbreviation for *bei*, or "by"
bei	"by", or "near"
Bez.	Administrative district
Dienstglas	"Service Glass"; binoculars
DIN	*Deutsche Industrie- Norm*; or "German Industrial Standards"
D.R.G.M.	*Deutsches Reichs Gebrauchsmuster*, or "German National Registered Design"
D.R.P.	*Deutsches Reichs Patent* (in this context; in others, *Deutsche Reichs Post*)
Dr.-Ing.	Senior engineer
Ersatz	Replacement; substitute

Flash	Excess material ejected out of the side of the mold, the seam, sometimes has to cleaned up.
Gau	Region, or state
Gebr.	Abbreviation for *Gebrüder*; or "brother(s)"
Geschichtete	Layered
Gesellschaft	Society, association, company
G.m.b.H.	*Gesellschaft mit beschränkter Haftung*, or "Limited Company"
Heeres-Abnahmewesen	Army Inspection Office
Heereswaffenamt	Army Weapons Office
HWaA	Abbreviation for the *Heereswaffenamt* or "Army Weapons Office"
i.	Abbreviation for "*im*", or "in"
Ing.	Abbreviation for "*Ingenieur*" or engineer
Kasein	Casein
K.G.	Abbreviation for *Kommanditgesellschaft* or "Limited Partnership"
Kriegsmarine	Navy
Krs.	*Kreis*; or district
Kunstharz	Synthetic resin material
Kunstharzpreßholz	Synthetic resin bonded, pressed wood
Kunsthorn	German generic for Casein
Kunstoff(e)	Plastic(s)
Luftwaffe	German Air Force

Mottle(d)	Appearance, due to purposeful incomplete blending of the pigmented molding compounds, used to simulate wood grain, marble, etc.
MPD	Abbreviation for *Staatliches Materialprüfungsamt*; or, "State Materials Testing Office", located at Berlin-Dahlem
Ofr.	Oberfranken
OKH	Abbreviation for *Oberkommando des Heer*, Army HQ
OKW	*Oberkommando der Wehrmacht*, roughly equivalent to the US Joint Chiefs of Staff/DoD
Patrone(n)	Cartridge(s)
Phenolharz-Kunststoffe	German generic for phenolic-formaldehyde polymer plastics
Pr.	Abbreviation for Preußen, or Prussia
Preßschichtholz	Pressed laminated wood
Preßstoff(e)	Pressing material(s)
Ratgeber	Advisor
RBNr.	Abbreviation for *Reichs-Betrieb Nummer*
RDRI	Abbreviation for the *Fachgruppe deutsche Rundfunkapparate-Industrie*, a specialized trade association of radio accessories (particularly speakers) manufacturers.
Reichs-Betrieb Nummer	"National Contractor Number", roughly translated
Rhld.	Rheinland
Sa.	Sachsen, or Saxony
Schußwaffenfabrik	Firearms manufacturer
Staatliches Materialprüfungsamt	or "State Materials Testing Office", located in Dahlem near Berlin (now a district of Berlin itself), abbreviated *MPD*
Thür.	Thüringen, or Thuringia

TL	Abbreviation for *Technische Lieferbedingungen,* or "Technical Delivery Conditions"
™	Trademark
Über	Over, above
VDE	*Verband Deutscher Elektrotechniker,* or "German Federation of Electrical Engineers"
Vogtl.	Vogtland or Western Saxony
WaA	Abbreviation for *Waffenamt*
Waffenamt	Weapons Office
WaPrüf	Abbreviation for *Waffenamt Prüfwesen,* or "Office of Weapons Proof and Development"
WDRI	Abbreviation for *Wirtschaftsstelle der deutschen Rundfunkindustrie,* or roughly translated, "Economic Association of German Radio Industry"
Wehrmacht	Military
Werkstoff	Material
Westf.	Westfalen, or Westphalia
Württ.	Württemberg
Zellhorn	Cellulose polymer plastic, such as Celluloid
Zelluloid	Celluloid
Zeugamt (Zeugämter)	Arsenal(s)/Depot(s)
Zünder	Fuse
Zünderbüchse(n)	Fuse Container(s)

Bibliography

Backbone of the Wehrmacht, by Richard Law, Collector Grade Publications, Cobourg, Ontario, Canada, 1991

Bakelite Technic: Molding Materials, published by the Bakelite Corporation, New York, 1928

Bekanntmachung über Kunstharz-Preßmassen für typisierte und überwachte Preßstoffe, an article appearing in the journal *Kunststoffe, Bd. 30, Heft 3*, 1940

Belgian Plastics Industry, CIOS XVI-4, 5, 8, SHAEF, 1945

Bezeichnungen für Kunststoffe im heutigen Deutsch, by Gerhard Voigt, Helmut Buske Verlag, Hamburg, 1982

Desperate Measures: The Last-Ditch Rifles of the Nazi Volkssturm, by W. Darrin Weaver, Collector Grade Publications, Cobourg, Ont., Canada, 2005

Dr. F. Raschig, G.m.b.H, Chemische Fabrik, Ludwigshafen, BIOS Final Report No. 507, HM Stationary Office, London, 1947

Dynamit A.G., Troisdorf, CIOS XXXI-3, SHAEF, 1945

Formmassen Lieferprogramm, company prospectus published by Bakelite A.G., Iserlohn, Germany, July 2003

German 7.9-mm Dual Purpose Machine Gun MG34, TM-E9-206A, US War Department Technical Manual dated April 13, 1943

German Army Handbook, TM-E-30-451, US Army, March 1, 1945

German Casein Plastics Industry, BIOS Final Report No. 282, HM Stationary Office, London, 1946

German Mauser Rifle, Karb. 98-K, w/Plastic Stock and Handguard, ETO Ordnance Technical Report No. 111, US Army, 17 January 1945

German Military Letter Codes, by John Walter, Small-Arms Research Publications, East Sussex, 1996

German Plastics Practice, Office of the Quartermaster General, Research and Development Branch, US Departments of the Army and Commerce, 1945

Hitler's Garands: German Self-loading Rifles of WWII, By W. Darrin Weaver, Collector Grade Publications, Cobourg, Ontario, 2001

Infiltration, by Albert Speer, MacMillan Publishing Co., Inc., New York, NY, 1981

Investigation of German Plastics Plants, CIOS XXIX-62, SHAEF, 1945

Investigation of German Plastics Plants: Part IV, BIOS Final Report No. 433, HM Stationary Office, London, 1946

Kunstoffe: Zeitschrift für Erzeugung und Verwendung veredelter oder chemish hergesteller Stoffe, I. Jahrgang, compiled by Dr. Richard Escales, J.F. Lehmann's Verlag, München, 1911

MG34-MG42 German Universal Machineguns, by Folke Myrvang, Collector Grade Publications, Cobourg, Ontario, 2002

Plastics Technical Dictionary, by A.M. Wittfoht, Carl Hanser Verlag, Munich, 1992

Rauch Guide to the U.S. Plastics Industry, by the staff of Impact Marketing Consultants, Inc., Manchester Center, VT, 2000

Staatlisches Material Prüfungsamt Unter den Eichen 86-87, Berlin-Dahlem, BIOS Final Report No. 525, British MoD, 1946

Uniforms and Traditions of the German Army, Vol. 3, by John R. Angolia, R, James Bender Publishing, San Jose, CA, 1992.

Webster's II New Riverside University Dictionary, Houghton Mifflin Company, Boston, MA, 1984

Werkstoff Ratgeber, by *Dr.-Ing* Herwarth v. Renesse, Buchverlag W. Girardet, Essen, 1943

Unpublished works and Correspondence:

DeSpain, Leon: email correspondence with the author dated October 11, 2006 and February 1, 2007.
Finke, Ulrich: email correspondence with the author dated October 27, 2006.
Heidler, Michael: email correspondence with the author dated October 10, October 26, December 17, 2006.
Huddle, Kenneth: email correspondence with the author dated September 4, 2006 and January 5, 2007, letter to author dated October 26, 2006.
Marschall, Dieter: email correspondence with the author dated October 29, 2004.
Sturgess, Geoffrey: email correspondence to the author dated June 9, 2006.
Yelton, David: email correspondence with the author dated July 31, 2006.

Notes

1   *Webster's II New Riverside University Dictionary*, Pg. 900.

2   Though it tends to "amber" with age.

3   In 1939, Baekeland's General Bakelite Corporation was merged with the Union Carbide and Carbon Corporation (later named the Union Carbide Corporation).

4   See *Bekanntmachung über Kunstharz-Preßmassen für typisierte und überwachte Preßstoffe*, listed by firm.

5   See *Werkstoff Ratgeber*, pg. 40.

6   As of 2003, "Bakelite" was a registered brand of the Georgia Pacific Corporation in the US and a registered brand of Bakelite AG of Iserlohn, Germany, in 50 other countries.

7   See *Staatlisches Materialprüfungsamt Unter den Eichen 86-87, Berlin-Dahlem*, pg.1.

8   See *Werkstoff Ratgeber*, pg. 37.

9   v. Renesse: *Werkstoff Ratgeber*, pgs 40-41.

10   v. Renesse: *Werkstoff Ratgeber*, pgs. 39-40.

11   See Appendix II.

12   v. Renesse: *Werkstoff Ratgeber*, pgs. 39-40.

13   Some of these designations are still used in the German thermoset and thermoplastics industry, though the meaning of the letter and numerals have in most cases changed since the war. For instance "Z3" =cylindrical granulates of approximately 6 mm diameter, instead of "phenolic resin with cellulose filler".

14   *German Plastics Practice,* pg. 4.

15   See *Belgian Plastics Industry,* CIOS XVI-4, 5, 8, pg. 10.

16   Hermann Staudinger won the Nobel Prize for Chemistry in 1953 for his research.

17 See *German Casein Plastics Industry*, BIOS Final Report No. 282, pg. 1.

18 See Speer, Albert: *Infiltration*, pg. 138.

19 *German Plastics Practice*, pg. 243.

20 Except those subject to repeated stress, shock and impact, such as tent stakes.

21 "Jüttner" refers to *SS-Gruppenführer* Hans Jüttner, who headed the office responsible for arming the *SS* divisions.

22 Speer, Albert: *Infiltration*, pg. 138.

23 See *Belgian Plastics Industry,* CIOS XVI-4, 5, 8, pg. 5.

24 See *Belgian Plastics Industry,* CIOS XVI-4, 5, 8, pg. 6.

25 See German Plastics Practice, pg. 239.

26 See *German Casein Plastics Industry*, BIOS Final Report No. 282, pgs. 16-20.

27 *Dynamit A.G., Troisdorf,* CIOS XXXI-3, pg. 3.

28 See *Dynamit A.G.,* CIOS XXI-3, pgs. 3-5.

29 *Dynamit A.G., Troisdorf,* CIOS XXXI-3, pg. 6.

30 See *Belgian Plastics Industry,* CIOS XVI-4, 5, 8, pg. 5.

31 v. Renesse*: Werkstoff Ratgeber*, pg. 39.

32 It should be noted though that there are sequential gaps in the *Kunststoffe* number list, many of which correspond to numbers known to have been used by small arms makers. But, then again, there are gaps that do not correspond with any known small arms or ammunition makers.

33 *See* Marschall, Dieter: email correspondence with the author dated October 29, 2004.

34 See Walter, John: *German Military Letter Codes*, pgs. 5-6.

35 This includes Austria, Belgium, France, Holland Poland and Czechoslovakia; but not the Sudetenland, which the MPD considered part of Germany "proper".

36 Branches of the *Luftwaffe and Marine* were also established to inspect and accept their weapons and equipment.

37 Established in 1940 and initially headed by *Reichsminister* Dr. Todt and later *Reichsminister* Albert Speer.

38 Law, Richard: *Backbone of the Wehrmacht*. See chart pg. 154.

39 Ibid, pg. 67.

40 At it's peak, the *Abnahme* may have numbered over 25,000 inspectors. As the war progressed, a greater percentage of these inspectors were female, allowing more German males to serve in the *Wehrmacht*.

41 See Weaver, W. Darrin: *Hitler's Garands: German Self-Loading Rifles of WWII*, pgs. 9-10.

42 See v. Renesse: *Werkstoff Ratgeber*, pgs. 40-41.

43 See *Uniforms and Traditions of the German Army*, Vol.3, p.129.

44 See *Belgian Plastics Industry*, CIOS XVI-4, 5, 8, pg. 10.

45 See Heidler, Michael: email correspondence with the author dated October 10, 2006.

46 See Law, Richard: *Backbone of the Wehrmacht*, pg. 278,

47 See Technical Intelligence Report No. 111, *German Mauser Rifle, Karb. 98-K, w/plastic stock and handguard*, pg. 1.

48 IG Farben was a huge conglomerate of industrial cartels and eventually had a controlling interest in over 380 German and over 500 foreign firms. During the war, I.G. Farben produced all of Germany's synthetic rubber and lubricating oils, and a large portion of the country's fuel, paints, adhesives, steel, light metals, explosives, plastics and chemicals. In addition to its infamous *Zyklon-B*, I.G. Farben developed and patented polyurethane and polystyrene, and the nerve agent, *Tabun*; after the war the founding companies were again separated by the Allies and remain in business as large corporations of course today.

49 See *Belgian Plastics Industry,* CIOS XVI-4, 5, 8, pg. 6.

50 See *Dynamit A.G.,* CIOS XXI-3, pg. 3.

Notes

51 See *German Plastics Practice*, pg. 1.

52 *German Casein Plastics Industry*, BIOS Final Report No. 282, pgs. 2 and 16.

53 See *German Plastics Practice*, pg. 1.

54 *German Plastics Practice*, pg. 1.

55 *German Plastics Practice*, pg. 1.

56 *Belgian Plastics Industry*, CIOS XVI-4, 5, 8, pg. 12.

57 Synthetic resin molding compounds in the sense of this announcement are hardenable products to be molded warm (on the basis of phenol- or urea resins) as well as non-hardenable injection molding compounds (Type A) which are introduced into commerce as unshaped, semi-finished items. Synthetic resin moldings are the shaped bodies manufactured in the molding- or injection molding process; see also conceptual explanations in VDE 0320/1939 "Guidelines for Non-ceramic, Rubber-free Insulating Materials."

58 See the following "Announcement Concerning Standardized and Inspected Moldings," Tables I and III.

59 See "Standardizing of Rubber-free, Non-ceramic Insulating Moldings," Plastics Vol. 27 (1937) p. 30; Plastic Molding Compounds Vol. 7 (1937) p. 339; ETZ Vol. 58 (1937) p. 1254

60 Moldings in the sense of this announcement are the materials listed in the "Standardization of Rubber-free, Non-ceramic Insulating Moldings" (see Note 2) which are manufactured as shaped bodies from molding- or injection compounds in the molding, injection molding or spray process.

61 See *Plastics* Vol. 27 (1937) p. 330; Plastic compounds Vol. 7 (1937) p. 339; ETZ 58 (1937) p. 1254.

62 Likely a subsidiary of 38, hence the suffix.

63 See Heidler, Michael: email correspondence with author dated October 26, 2006 and Huddle, Kenneth: notes from proof copies.

About the Author:

W. Darrin Weaver MPA PA-C (Master of Physician Assistant Studies) raises cattle and provides primary care services to former servicemen at the Central Texas US Department of Veteran's Affairs. Darrin served in Berlin as an infantry medic from 1987-1992, was commissioned after PA residency in 1994 and since leaving active duty in 1999 has continued to serve in the Texas Army National Guard as a field-grade medical officer. Cited for marksmanship by both the US Army and German *Bundeswehr*, Darrin's other published works include the 2001 *Hitler's Garands: German Self-Loading Rifles of WWII* and 2005 *Desperate Measures: Last-Ditch Weapons of the Nazi Volkssturm*.